MOODY

MOODY

A WOMAN'S 21ST-CENTURY
HORMONE GUIDE

AMY THOMSON

 SQUARE PEG

1 3 5 7 9 10 8 6 4 2

Square Peg, an imprint of Vintage
20 Vauxhall Bridge Road,
London SW1V 2SA

Square Peg is part of the Penguin Random House group of companies
whose addresses can be found at global.penguinrandomhouse.com

 Penguin
Random House
UK

First published by Square Peg in 2021

Penguin.co.uk/vintage

A CIP catalogue record for this book is available from
the British Library

Trade Paperback ISBN 9781529110333

Typeset in 9.4/13 pt Change
by Integra Software Services Pvt. Ltd, Pondicherry

Printed and bound in Great Britain by Clays Ltd, Elcograf S.p.A.

The authorised representative in the EEA is Penguin Random House Ireland,
Morrison Chambers, 32 Nassau Street, Dublin D02 YH68.

Penguin Random House is committed to a sustainable future for
our business, our readers and our planet. This book is made from
Forest Stewardship Council® certified paper.

This is dedicated to all the people in my life and work who've supported me, through good times and bad, you all know who you are. Life is a team sport.

MEDICAL DISCLAIMER

INTRODUCTION

I want to share a secret with you, a secret that is already inside you. Once you understand it fully, it has the capacity to unlock untold potential. It is the science of yourself, of your own hormones. You might think you possess this knowledge already, but I can promise you there is so much more to learn; and when you have, you will be able to harness it forever.

Until a few years ago, hormones were not something I had ever paid much attention to; they were just something I'd learnt about briefly in school and part of the monthly hassle of bleeding for five days. I simply did not know how important hormones are in connecting our minds and our bodies. Then one day my periods stopped, and what had been a gross inconvenience each month became a barometer for my body's health and mental happiness.

The outcome of my own hormonal burnout was driven by stress. My body started to show signs, but I had not been trained to know what to look for. I can't identify the exact moment this happened, but the most significant sign was my periods stopping. My body bloated up and my mind was racked with chronic anxiety. It was the shock that I needed in order to listen to what my body had been desperately trying to tell me: *slow down; we're about to hit a wall.*

I was twenty-four when I had started a business that connected marketing with a new generation of social networks. I was young, but at the centre of a new digital decade, and fiercely ambitious. However, for all my work drive, I had been living clumsily within my own skin for my whole life. There came a point where my ambition, and my lack of understanding about stress, hormones and balance, meant I pushed my body too far, to the point of burnout.

It took a while to piece myself back together. My new focus became trying to understand how my body worked, to establish why and how I had pushed myself so far. I consulted every kind of doctor I could think of, as well as nutritionists, personal trainers and endocrinologists. An endocrinologist was not a doctor I had ever heard of until my hormones went haywire. We might be familiar with gynaecologists, a concept most of us know about from films or shows like *Sex and the City*, but it took me a while to realise my hormones had their own specialists too. All of these experts encouraged me to keep journals about my health, but none of them seemed to put the pieces together for me. In fact, in the early days, none of them even really connected my lack of periods with stress. However, it was these analogue journals that planted the seed which would eventually become the foundations for Moody Month, the app I went on to create to help women track their cycles, moods and symptoms.

I'm obsessed with research, science and data in all areas of my life, so I started by treating myself like a study: plotting my cycle on a daily basis, and also charting how I felt, the symptoms I experienced and what was going on around me. I read widely and consulted some of the best doctors and women's health practitioners and became fascinated by the science. I fangirled medics such as Dr Jen Gunter, Dr Angela Saini and Dr Alicia Viti, in the hormonal equivalent of obsessing over the music of the Beatles, Backstreet Boys or One Direction, subject to your age and hairstyle preference. I couldn't stop talking about it. I was the girl at every party, wedding or dinner going on about hormones. One of my friends even started calling me, charmingly, MenstruAmy.

With all this personal data, I was able to connect the dots between my peaks of stress and anxiety to when my cycle would become most disrupted. I began to see how the external patterns of my world were affecting the internal patterns of my hormones and health. What happens to me when I don't listen to my body is that my periods stop. My body clenches up and I become tight. I constantly feel like I'm just about to drop from the highest point on a white-knuckle roller coaster. The longer my body is in this state of tension, the more manic and frantic my mental state becomes. In hindsight it's laughable how little I knew about my own body. My entry-level understanding of hormones was that they were connected to my period. That's the equivalent of only knowing how to say 'please' and 'thank you' in a new foreign language – a good start, but pretty basic. So many women I knew, including myself, seemed so uncertain of exactly why or how they work.

What I know now is that there is no perfect pill to solve all your problems, and that there is no singular doctor or expert who will help you understand yourself or your cycles. Whatever practice or support you look to when your body is out of whack, the first thing any doctor, practitioner or healer will ask you is: 'How regularly are you experiencing these symptoms?' They want to see if there is a pattern in the symptoms, in moods, in experiences.

Humans have looked for patterns in life for millennia. Patterns are key to almost everything, from nature, science and politics to money, art, culture and even sex. It is how we make sense of information. When something doesn't make sense to us, it is often that we simply haven't been able to connect the dots at that point in time – have you ever had a moment, looking back, when it all suddenly clicks into place, and you finally see something that was there all along? Your brain just solved the pattern.

That was how I felt when I started to log and study my own symptoms and moods and understand the hormones behind them. How they gave meaning to the often random feelings I had or symptoms I felt. Frankly it was a huge relief – my 'aha' moment.

Have you ever had sex with someone and felt awful after? Have you ever overeaten and slipped into a food coma? Have you ever had those days where, no matter how hard you try, you cannot switch your focus or memory on and feel like you might have lost your touch? I have done all these things, sometimes on a bimonthly basis. What I'd never been told before was that there are patterns to these moods and symptoms. You can look out for these patterns and use them to help you steer your mind and body. Our decisions are often driven both emotionally and physically by these patterns, which are themselves caused by the hormones inside us. Rather than feeling guilty about the inevitable highs and lows of life, when you understand how hormones could be part of this experience, you forgive yourself for actions outside your control and start to build better and healthier processes and routines. That is why I feel so compelled to share the research and science behind how our hormones work for 21st-century survival. Throughout this book I will unpack the systems and cycles inside our bodies that can be an invisible but powerful force.

What I came to learn about my own cycles and patterns was how to listen, learn and then optimise my life around them. By accessing the science, it made me realise it wasn't always my fault that I acted impulsively, felt sad or slept with someone my logical mind said I shouldn't. When you get some of the background to hormones and how they work, you can be more aware of what lies behind your emotions, and it might even lead you to be more constructive in certain decisions. It is important to be clear that hormones don't control us, we still have free will, but they're an important variable to consider in our daily decision-making.

This deep dive into the science has taught me the power that hormones hold in my ability to find a work–life balance. Emotional intelligence and energy management are two superpowers everyone should be armed with for success in today's ever-changing world, and both can be supercharged by understanding the effects of hormones. The key for me was energy management. Unlocking the logic of how my body works has given me some incredible powers for focus and productivity in my day-to-day, even arguably the power to get funding for what would eventually become an app based on this research.

There was one moment that changed everything for me, when Lola Ross, my nutritionist, described the fundamental relationship between hormonal patterns and mental and physical well-being. How these patterns of hormones are the cornerstones for understanding why and how our bodies work. 'For many women I work with, this connection is surprising new knowledge,' says Lola, a Moody co-founder and registered nutritionist. 'This is why hormones need more airtime. When it comes to health, like so many things, knowledge is power. The more we know about the role of hormones, the interconnectedness of the endocrine system and its role in health and well-being, and the better we understand how hormones can be both positively and negatively influenced through our environment, diet and lifestyle, the more power we have to better support our health potential.'

In 2017, after a year of paper journals, experts and self-care as the solution to stress, I realised not only did I need to change my life, but find a way of sharing all of the science and the superpowers. I sold the business that I'd worked so hard to build – but which had burnt me out in the process – and went deep into the academic side of hormonal research. I was in a unique position, between being a civilian with a personal fascination, but also having access to world-class doctors in endocrinology, psychology and neuroscience, along with the health and well-being experts I had worked with to put myself back together. I was on a mission to make the conversation about hormones cool, accessible and an open source to support daily and weekly wellness routines for as many women as possible. This would be the start of my journey.

I used the very science that I was being told was niche, to keep my mental and physical health on track and maintain the momentum I needed to raise the capital and build the tech. I began putting together a team of female engineers and data analysts to build technology that allowed women to log their menstrual cycles, symptoms and moods. It is all about daily and weekly pattern recognition – a weather forecast for your moods and hormones. The aim is to help individual women make sense of their body's unique hormonal code, to support long-term well-being and decision-making. I've heard back from thousands of women who use the app, and they say it's been transformative; that they wished they'd had this knowledge before. It's our collective job to spread this awareness further.

We chose the name Moody as a way to reclaim the word. It's not a badge of shame caused by our hormones, but a trophy of our humanity and emotional intelligence. We are all moody; it's what makes us human. Without moods and hormones, you're numb; you might as well be a robot.

The purpose and focus of understanding our cycles is not to control and suppress them; rather, it is to encourage us to listen and learn so we're able to support ourselves in daily life. This has been my greatest lesson and my aim is to inspire you as the reader to trust your own body, listen to its patterns and, in turn, share your story to help other women.

I'm passionate about this subject because I know that under-standing your body, self and hormones gives you so much power. It is also important to highlight that I am one kind of woman: I am white and middle class. What has also been very apparent in the research I have done is just how under-represented other voices are in this space, specifically women of colour and the trans community. While I cannot appropriately represent other women's experience in detail, I work with and have aimed to share throughout some of these voices.

One of the best pieces of advice I have ever been given was that you are the most qualified person to understand yourself. Tuning in to your body and spending some time listening and learning from what feels good and what feels bad is an essential part of our self-discovery. Everyone's journey is different. You need time to learn any new language and learning the language of your body is essential. My aim with this book, as with the app, is simple: I want to help women – all women – become fluent in the language of themselves and their bodies.

HORMONE
HEADLINERS

Imagine if someone had sat you down when you were a little girl and told you that you had superpowers. That those super-powers – hormones – are responsible for all those big feelings that make you act in certain ways, and that they help shape your life through the decisions you make. That they will affect everything: from your body confidence (or lack of it) to your memory, energy and productivity, your emotional highs and lows, to the relationships you choose, and the people you fancy. That they are more important than the subjects you choose to study at school and the career you end up in – because they influence all those decisions. But, crucially, that you aren't hostage to these little chemical messengers. That you can listen to them, nurture them, and they in turn will help you live your best version of you. Wouldn't you want to learn more?

Every emotional action and decision has a chemical pathway, from brain to body, that is built up throughout your life. Every action has a hormonal memory and over time these become more prominent. These pathways can be good and bad: seeing a birthday cake, for example, will for many of us trigger a hormonal memory path, linking back to all the happy memories of previous birthdays and eating sweet treats. This is a reward pathway, and as we get older and become more exposed to stress, we can often crave these rewards more regularly. The trick is to build rewards into your life that can help support stress, but don't have high sugars and fats. Building a treasure chest of ways to help self-soothe in the inevitable low moments.

Throughout this book, I will give some of the chemical backstory for how and why we need these physical and mental rewards and how they can not only make us feel better in the moment, but support long-term health and wellness.

Over time, good and bad chemical pathways begin to become more set, so helping unlock the happy hormones with rewards is key and this is the science that lies behind self-care. Your body takes in stimuli from the outside and turns them into a response; when you tune in you can hear the signals and decode or understand that response. These internal signals determine when you feel hungry, happy, sad, sexy, sluggish or motivated. These moods, emotions and symptoms are the language of our body that no one has really thought to share with us.

What most people know of hormones comes from basic biology in school. That was certainly true for me. We are taught that estrogen is connected to women, periods and pregnancy and that testosterone is connected to men. From the various contraceptives I tried and tested, I went on to learn first-hand that hormones could make me moody. In my twenties, I learnt a little about serotonin, but mostly due to partying and the horrific emotional low you hit on the morning after a big night out. These (kind of basic) 101 introductions to hormones are commonplace for most of us.

What I would come to discover is less well known: not just how the hormones inside me worked, but how happy hormones are actually internal medicine and how – when I understood them – I could self-administer them when needed. I wish that someone had told me all about this earlier. It could have saved a lot of confusion.

When burnout came for me, and my periods completely stopped, the root cause, I would later discover, was stress. By that point I was twenty-nine and had never been told that stress could stop my periods. No one had ever explained to me that stress was not just a 'feeling'. It was a chemical chain reaction triggered by external stress stimuli. The hormonal reaction to stress – overproduction of adrenaline, cortisol and noradrenaline – is your body's internal caffeine, designed to push you through. If you've ever had one too many cups of coffee you'll know it's not a good idea to spend your life running on it, as it can make you painfully anxious. Similarly, stress hormones are your body's response trigger to push you through danger. Have you ever felt the rush from adrenaline? It makes you hyperalert and your senses rush and tingle from the world around you. It's why it can often be very addictive. Living day-to-day with this powerful compound of internal fight or flight leads eventually to a knock-on effect and crash.

Hormones don't benchmark themselves against a cultural lens of what should or shouldn't be stressful or dangerous; they simply respond based on what your brain reports back as stressful or dangerous. Our brains have evolved into supercomputers: we deal with all kinds of complex emotional and environmental factors in the twenty-first century, but our hormones still serve as the internal response system to regulate our body's rhythms.

YOUR DAILY DOSE

Your body has an internal pharmacy of chemicals, being released in doses that are determined by your glands. Alongside your stress hormones, you also have your 'happy hormones', a combination of hormones and neurotransmitters which are chemical signalling molecules. They are essentially a set of superpowers you can learn to work with effectively and self-administer through wellness routines. They don't just make you feel good; they can help guide you through your day-to-day. These hormones have the power to make us all feel strong, seductive and supercharged. They can support energy, focus and memory. What do they consist of? Here's the rundown:

Dopamine is your happy reward drug and functions as both a hormone and neurotransmitter. It's administered by your brain and connects with your body to tell it that something feels good and is worth doing again.

Oxytocin is known as the 'cuddle' hormone, linked to our desire for bonding. It is powerfully released through human interaction and touch.

Serotonin (whose chemical name is 5-hydroxytryptamine or 5-HT) is your mood-boosting hormone and neurotransmitter. It is the chemical that evokes the emotion of happiness and well-being. It is also integral to sleep pathways.

Endorphins are a collection of neurotransmission chemicals that are released as your internal pain relief; they work like morphine inside. They are what drives the happy high after a hard workout. They also stimulate other hormones inside your system.

It is the combination of these chemicals that is at the core of our quest in life to feel happy and healthy. They are your internal 'love drugs' or **DOSE**.

We can proactively access our daily DOSE in all sorts of ways – from seeing friends and family, eating foods we love, exercise, or self-care such as having a massage, facial or a long bath. Following childbirth, women are administered by their own bodies one of the strongest injections of DOSE – an antidote to the pain, but also an influx of important chemicals that drive the early bonding experience with their baby. DOSE is not just about bonding with other people; it is also important for how we self-heal and bond with ourselves. It is our body's antidote to physical and mental pain. DOSE is your body's power pack – a solution to sadness and stress. We are not designed to feel sad or sit in sustained pain, but it is an emotional state that everyone experiences in life. We know when these happy hormones kick in, as our bodies and minds become clearer, focused and calm.

This happy cocktail is the reason wellness works and makes us feel good. There is even an entire content platform dedicated to this very topic, DOSE, which Shara Tochia and Hettie Holmes co-founded after realising that the fitness industry revolved around weight loss or how to get abs in six days. What was missing was the fact that fitness and wellness are more powerful as tools for happiness than short-term weight loss or training goals. They describe their audience as 'healthy hedonists', which I think is a great way to be: a balance between being healthy and having a social life. As Shara puts it, 'We didn't care about weight loss, we wanted to build content that talked to us as normal women and not athletes. We wanted to read articles that didn't make us feel guilty for wanting to work out and for also having rewards in our lives. We wanted to both feel well and go out for a pizza occasionally.'

So DOSE is your body's internal injection of relief – but this also means you can get 'high on your own supply'. Sex, for instance, is a cascade of hormones, started by the first touch and eliciting some of the highest possible dosages of DOSE. The most powerful part is that when we feel pleasure from another person our bodies can chemically bond through the hormone oxytocin. But it is possible for this to tip over into something else: with sex, the highs from these chemicals can lead to 'the chase' – and even addiction – when you hunt or want more and more of that feeling.

When we are in a state of addiction, our bodies and minds crave and search for the happy hormones from stimulants. Whether the stimulant is drugs, alcohol or sex, they all can provide quick happy hormonal hits. But longer-term addictive behaviours and triggers mean the effects of these stimulants begin to inhibit your brain and body's ability to produce the very hormones and internal happy chemicals you crave. This creates the cycle of addiction, as your body then becomes dependent on external stimulants to produce happy hormones as it cannot produce them itself. (Technology can also sit within this category of reliant stimulants and research into the effects of addictive technology habits in gaming is beginning to shed some light on just how powerful the effect can be.)

So DOSE is necessary for our well-being, but our search for it has to be balanced. To ensure you are able to keep steadily replenishing your internal supply of DOSE, your body needs sleep, food, water and human contact. These needs haven't changed for millennia. What has changed in the modern age is the introduction of career goals, financial expectations, fast food, fad diets and Tinder. We are not any longer trying to escape sabre-toothed tigers, but there are still some pretty mammoth milestones in the 21st-century technological era that affect our ability to access happy hormones.

THE HORMONAL DANCE

There is no way of controlling all elements of your life or the environment around you. You still have choices, and employing the lessons you have learnt from life experience *alongside* the science of your hormonal patterns is essential. What I can share is the current knowledge and some of the common experiences women often dismiss as character flaws or failures – when actually they're hormonal. I discovered so much about these systems, by finding myself in a very unique position triggered by a personal hormonal burnout, a cultural shift in the world of women's health, and a global realisation that mental health is not a nice-to-have, but essential for survival.

Hormones are so much more than your period. Your hormonal self-discovery should start with the endocrine system – the collection of glands that produce and secrete hormones, and where they are across your body; something few of us know about. Here is a quick diagram to show you where each of these power packs are positioned. Each of them operates like a valve of signals and signposts to keep your whole body regulated, your organs functioning and producing energy for the day and recovery through sleep at night.

THE FEMALE
ENDOCRINE SYSTEM

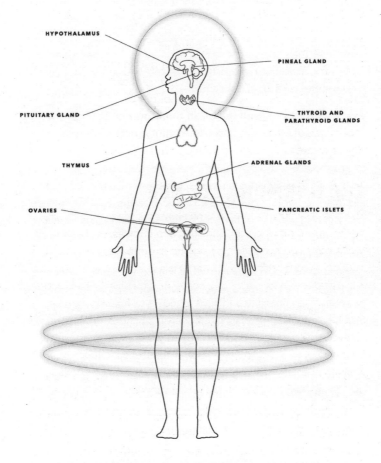

HYPOTHALAMUS

PINEAL GLAND

PITUITARY GLAND

THYROID AND
PARATHYROID GLANDS

THYMUS

ADRENAL GLANDS

OVARIES

PANCREATIC ISLETS

I will come on to the detail of each gland and hormones
associated with it later in the book. You may have noticed
some of these glands when you are run-down or suffering
from fatigue, as we often ache in those spots. This is the body
working overtime to support your natural immunity or trying to

provide extra energy to push through the low. Each gland has a specific set of chemicals that are released as a response or solution to an internal or external trigger. Your glands are your internal pharmacists, distributing and prescribing what chemicals your body needs and when. It is this distribution balance that ultimately defines whether your body is feeling happy and healthy or slow and sad. However, when they are working overtime, they can become less efficient in their decisions and their production can become disrupted causing a change in you mental and physical clarity.

Once you start to read the good and bad signs of hormones, it unlocks your intuition and allows you to navigate this sixth sense with much more logic. Intuition is often seen as magical or mystical, but the power seems to be much more hormonal to me. Dr Tara Swart is a neuroscientist, and author of *The Source*. Her work looks at how breakthroughs in neuroscience are proving we can actively alter the way our brains function, by understanding the systems inside ourselves. Her insights on neuroscience echo and underline how our bodies – from brain to hormones – carry an imprint of memory: 'Our experiences, and especially those experiences we repeat more frequently, are all stored inside our brains. We don't remember every experience of our life, but they are stored within neurons in the brain stem, spinal cord and gut neurons. When we have a "gut feeling" it is based on pattern recognition, though you can't consciously recall it. This experience will have a subconscious effect on your neurons and thus create an "instinct" or reaction. Instincts are like neurological memory recall ... Let's take everyday experiences like getting into a new relationship, asking for a pay rise, or applying for a new job – when your brain is confronted with the feeling of threat. The amygdala and hippocampus – the seat of emotions and where emotion and memory are connected – are dredging up the most negative memories. Memories and past experience of stress therefore create a hormonal response that is not just from the present experience, but from past historic memories. This means that your gut instincts and chemical responses are a combination of the memory systems inside our body triggering hormones.'

In other words, there is a chemical pathway that is laid down through the course of our experience that has direct implications for our hormonal responses.

By knowing what chemicals are inside us, we can expand our emotional language and unlock our superpowers. On the following page is a simple key for some of the hormones I will unpack throughout the book.

These chemicals inside us have immediate effects but also long-term impacts on the wider hormonal cycles and organs. They can affect everything, from how well we metabolise the food we eat to whether we feel in the mood for sex and how well we sleep.

The effects of hormones don't just drive immediate impulses like tiredness or hunger; they can also affect appearance. The condition of our hair, skin and nails is determined by the balance of hormones inside the body. There are some days in your menstrual cycle when you literally glow from the inside out. These days are when your three menstrual power players – estrogen, testosterone and progesterone – are prepping your body for ovulation.

There are three major cycles that can be tuned in to barometers for your health and happiness, to support the balance of your internal pharmacists and their prescriptions. These are the chief functions for your body's ability to re-regulate your chemical balance each day, week, month and year. For most cisgender pre-menopausal women, these cycles are: menstruation, sleep and metabolism. Transgender women and men have the additional hormone cycles of transition. I have spent a lot of time working with and talking to the trans community about how we can also think about building tech to support

CHEMICAL KEYS TO WELL-BEING

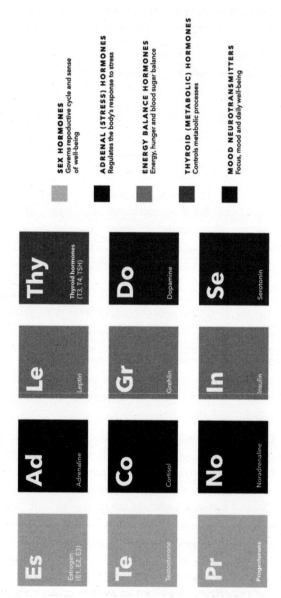

Es Estrogen (E1, E2, E3)	**Ad** Adrenaline	**Le** Leptin	**Thy** Thyroid hormones (T3, T4, TSH)
Te Testosterone	**Co** Cortisol	**Gr** Grehlin	**Do** Dopamine
Pr Progesterone	**No** Noradrenaline	**In** Insulin	**Se** Serotonin

SEX HORMONES
Governs reproductive cycle and sense of well-being

ADRENAL (STRESS) HORMONES
Regulates the body's response to stress

ENERGY BALANCE HORMONES
Energy, hunger and blood sugar balance

THYROID (METABOLIC) HORMONES
Controls metabolic processes

MOOD NEUROTRANSMITTERS
Focus, mood and daily well-being

these particular cycles. The hormone cycles cis women have can all be tracked, but no women and men need more support than those transitioning. I have included some of their experiences later in the book.

Certain medical conditions and interventions can also mean the loss of periods and ovarian cycles. Ellamae Fullalove is the founder of Va Va Womb, a stigma-breaking health community, and Mind Over MRKH to raise awareness of the condition MRKH (Mayer-Rokitansky-Küster-Hauser), after being diagnosed with it at sixteen. MRKH is a congenital disorder where women are born with forty-six XX chromosomes, functioning ovaries, but have an absent or incomplete vagina, no cervix and either an underdeveloped uterus or no uterus at all. Everything externally is unaffected so women with MRKH have a vulva, breasts and go through puberty. One key sign of MRKH type 1 is the absence of a bleed during the ovarian cycle. MRKH type 2 is more complex and can impact the kidneys, skeletal structure and/or hearing. The condition affects 1 in 5,000 women worldwide. Ellamae brings to the world an optimistic and positive view that we need to change how women regard their bodies, and not assign traditional norms or stereotypes associated with menstruation or fertility as the definition of femininity and womanhood. She is both an activist and an advocate for all women, with a belief in a more body-positive future. As she puts it, 'So many people with MRKH are afraid of platforms focused on women's health, as these platforms often focus on bleed and fertility, which can be alienating for those who don't bleed or who suffer from infertility. Platforms and brands often use the term "all women" when talking about fertility and bleeding, but this is not all women, as not all people who menstruate are women. The conversation around ovulation tracking is revolved a lot around fertility and getting pregnant which isn't inclusive of those doing it for reasons other than trying to conceive. There is so much confusion and misinformation around MRKH and a lot of our community didn't know they could track their ovulation

NON MRKH

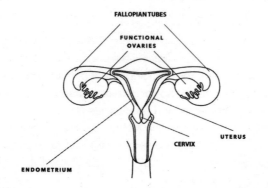

FALLOPIAN TUBES

FUNCTIONAL
OVARIES

UTERUS

CERVIX

ENDOMETRIUM

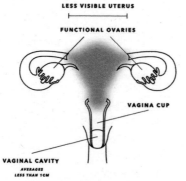

LESS VISIBLE UTERUS

FUNCTIONAL OVARIES

VAGINA CUP

VAGINAL CAVITY
*AVERAGES
LESS THAN 1CM*

MRKH

cycles and the hormones associated, as this is such new technology, or they haven't tried due to lack of resources. We need to be more inclusive of the language we use when talking about women and fertility. My femininity is not chromosomes, womb or bleeding; it comes from my decision to be a woman and identify as a woman.'

Hormones are a human reality – and no matter whether you identify as a woman, or trans, or non-binary, you have hormones and patterns that can be harnessed to help you understand your mental and physical health. Menstruation is one cycle that can be tuned in to, but all women are born with a unique pattern, body and rhythm: it's just about tuning in to yours.

Our hormones' cycles operate like a symphony: each has its own rhythm and they work around a 24-hour timing. This beat is our circadian rhythm, the body's 24-hour system that starts and resets. I will explain more on how and why these cycles and rhythms are so important to our happiness, and why they are crucial to our levels of DOSE, as we go through the book. These three symphonies work in harmony with each other, governed by the chemical signals that are hormones. When one of them is off-key, the whole band starts to fall apart. How to tell when this is happening to your body? You begin to see physical signs, from disrupted sleep to weight gain or loss, or irregular periods. Often, we plough on with our lives without listening properly to these warning cries for help from our bodies, trying to plug the gaps of our natural energy with sugar or caffeine, to give us a much-needed booster. But, as I found out myself the hard way, it rarely works in the long term and ultimately this fake energy will only get you so far. You have to look at the hormones within and work out where the imbalance is coming from. Our bodies are rhythmic, and if we are listening to the signals, they have ways of communicating with us when they're out of tune. I've used that knowledge to supercharge my own life in all its areas.

But what is most inspiring about this science of happy hormones is they offer an internal antidote to the inevitable sad or low times in life. By building a better relationship with the way I balanced my hormones, my periods came back and my immunity, energy levels and moods totally balanced. I stopped being permanently tired and catching colds. Driven by healthy and simple patterns in self-care, eating and knowing when to push myself or not do exercise, I was able to recover my equilibrium. There is no way of controlling what happens to your internal or external self, but there are ways to be more kind to both sides. It is personal pattern recognition: when you see a pattern that seems to go off-key, your body is trying to tell you to act or change course.

We all learn about our unique hormonal highs and lows by doing the things in life that have an impact on our happiness and sadness. In theory, we can use this knowledge to build better understanding of how to self-regulate. When we have been through years of hard-won experience, we begin to learn instinctively who we should or shouldn't sleep with, or how much wine will take us 'over the edge', or that if we forget to eat breakfast we will experience a sugar-low. If you pay attention to it, the chemical brew that is inside us all can also become part of your ability to consciously self-regulate. This is the point of sharing more of the science: as a more mindful way of accessing and understanding what makes us happy or sad, empowering us to make conscious choices about our patterns.

Understanding our patterns, pathways and hormonal reactions can stop us feeling guilty or questioning our very sense of self, because we can now see that some reactions are just chemical. I want to inspire people to think about the meaning that can be found between life experiences and the hormonal patterns at play.

We're living in an era where well-being is currency, but expectations are high. It is not realistic to spend every day doing two hours of yoga and eating only 'clean' food. For most of us, real life is not a picture-perfect Instagram feed of health and wellness; it is messy and sometimes you just need to eat cake. This culture of wellness as a way of life, especially in its online manifestations, can lead to a sense of failure if you are not living up to its expectations. This leads to more stress and anxiety, rather than inspiring long-term routines and rituals. Stress and anxiety disorders are an epidemic and finding long-term well-being and self-care routines can be an internal chemical support. Our hormones can be used to guide and navigate a more realistic personalised routine, if we know why and how they work. They govern everything about our bodies and minds, and are at work in every single one of our cells. We aren't used to talking about the hormones that produce the chemical reactions inside our bodies, nor the accompanying – and sometimes bewildering – feelings they concoct. We have to learn the hard way – by living through it. And if we don't get it right, if we ignore our hormones and the signals they try to send us – desperately, at times – then we can end up with a hormonal imbalance that spells serious trouble. It can be small-scale, such as disrupted sleep for a few nights, or it can be much more dramatic.

What shocked me, and is a common report from women we at Moody speak to with menstrual-related issues, is how long it took for anyone in the medical profession to tell me stress and my lack of period could be connected. There was a real sense of dismissal from male doctors that my period was at all linked to any of the other symptoms and the stress I was experiencing. This is when a friend gave me the book *Inferior* by the science journalist Angela Saini. In it Saini paints a disturbing picture of how deeply sexist notions are embedded into scientific research and the medical community – and how they are still being perpetuated.

I suddenly realised the reason why both medical research throughout history and the doctors I was currently working with had not been looking at the female experience of hormones. It wasn't because our hormones are not a brilliant source through which we can understand ourselves, but simply because women's health and hormones are governed by a male bias and no one thought they were worth focusing on. Caroline Criado-Perez investigates this brilliantly in her book *Invisible Women*. She writes about a cacophony of instances that have led to the world being governed by a male experience, from crash test dummies in cars to the seats of bicycles, but one that stood out for me was the lack of research into how medicines interact with women's fluctuating hormonal cycles: 'Menstrual-cycle impacts have so far been found for antipsychotics, antidepressants, antihistamines and antibiotic treatments as well as heart medication. Women are also more likely to experience drug-induced heart-rhythm abnormalities, and the risk is highest during the first half of a woman's cycle. This can, of course, be fatal.'

What was very apparent in the journey to build Moody was just how frustrating this is to see through so many women's experiences and how this has limited so much progress for equality. It is one of the reasons that sharing experiences and giving platforms to voices is so paramount to change. Saschan Fearon-Josephs founded the Womb Room, an online space for women to learn from other women suffering from reproductive-health issues. She works in particular with bringing the voices of women of colour to the fore, and has seen first-hand how they can find their symptoms being overlooked: 'One of the things I find consistently from the Black women I speak to is not being listened to first time around, and having to really fight and advocate their pain, but then being given very limited options for treatment. It has been a noticeable trend and something that is important to address. As a Black woman, you have to fight a lot and this is one area where being believed and supported is even more crucial.'

Similarly, Lola Adesioye speaks of the challenges Black women continue to face in the medical system. A race activist and journalist, she talks about the historic bias built into the training of doctors and in broader society, and how this poses profound difficulties for women of colour: 'What doctors and society often forget is we're all victims to fundamental attribution error. This is a principle from social psychology that means as humans we will always over-emphasise and lean into our own experiences as the dominant norm. These "norms" have become the fabric of society, as the politicians, doctors and thought leaders who defined our current social norms were mostly white men. Our society was born from colonialism. Anything pre-colonialism is often dismissed as primitive or basic. Therefore, colonial bias and scientific language has formed Western civilisation and the experiences we all have today with doctors and hospitals. This society is racist and sexist due to the bias that was built in from the fifteenth century. When you have under-represented groups or people talking about important topics such as hormones, the system is not set up to listen, as it is designed from white male norms, which we have all come to accept as normal. But they are not normal.'

The bias isn't always intentional, but the impact of bias does affect our experience of the world. It forces us to subconsciously and/or consciously dismiss our own experience and conform to someone else's opinion. We all need systems and structure to avoid chaos, but when your body has been dismissed for centuries as a vehicle, rather than an instrument to learn, it does make trusting yourself harder.

When my periods stopped and my body's physical reactions to burnout became unavoidable, I realised that no one had given me permission to listen to the patterns of my hormones or told me how much power they hold. The issue was I was waiting for someone else to tell me what to do and how to work my own body, but I needed to understand them myself. This then allowed me to make peace with the facts that our bodies evolve, change and are hugely impacted by the choices and environments we live in. Tuning in to the power of hormones, was tuning in to the intuition and baseline systems that make us human. Suddenly I realised my cycles were a lot more powerful than just for understanding when I was fertile or about to bleed. This science allowed me a new language to talk about my body and mind with everyone around me. I was given the keys to a behind-the-scenes investigation of my own body, which struck me as information that everyone should know, not just doctors or a few people who have privileged access.

I felt sad that I hadn't been told earlier just how helpful this would have been in navigating life, the world and my bur-geoning sense of self. Making hormones cool and accessible was an important mission, but not always an easy one. Men don't want to talk about hormones all day, and due to the amount of capital needed to launch a business focused on making this knowledge more accessible, I had to talk to hundreds of men with money. I was faced with some very real roadblocks from investors, who simply thought hormones were niche, even though they affect 100 per cent of the human population and for women are one of the best and most effective ways to organise and monitor your mental and physical well-being.

This book outlines why understanding our personal patterns can be so powerful. Its aim is to share the science and real-world experiences of how hormones can be harnessed and navigated by simply knowing how they work. Not just because the science is so compelling, but because when I began to look back at my life with this new knowledge, I started to understand patterns of pain and pleasure. The positive by-product of this awareness and pattern recognition is being able to see the funny side of how hormones make us do some pretty ridiculous things. It was almost as though my body gave me the ultimate DOSE from laughter, as proof I had healed.

Everyone's life is a tapestry of human interactions and sliding-doors moments. If you haven't seen the nineties classic film *Sliding Doors*, you should. The film is centred around the simple truth that in a split-second experience or interaction, your life pattern can change course, sometimes without you even knowing. What I want to share is not just how these patterns are emotional, but how they are hormonal too. What was your last sliding-doors moment? It could be a conversation about a career choice that led to leaving your job, or a chance Tube journey where you met the love of your life. These are the 'Moody' moments that make us who we are today, and craft some of what we may be tomorrow. My own Moody moments will be peppered throughout this book, from the shame of one-night stands to the guilt I built from cake addiction. Nowadays I can look back on these and laugh, giving me a shot of DOSE rather than piling on the shame or stress.

Like all information in the world, your body is about interpretation. The longer you live in your body, the more familiar you become with it. This is how you can support yourself in daily, weekly and long-term wellness that works for you and your unique pattern. Your body grows, it evolves, it breaks and changes course. But we can all be experts on our own chapters and stories.

This book is the start of a chain letter, from me to you, and you to the next woman. Because by understanding our own bodies and feeding that knowledge into the world, we give more knowledge and power to more women. Just by sharing how you feel and what you have experienced in your health and well-being, we can gain so much insight into a future for women, where women are listened to. So, consider this the permission you need to let rip and get to know your own mind, your own body, your hormones – and to spread the word. By the end of the book I hope you will be looking at yourself with a fresh set of eyes. How did your hormones make you?

TAKEAWAYS

- The happy hormones inside you are DOSE: dopamine, oxytocin, serotonin and endorphins.

- Your endocrine system is a series of glands across your body that excrete hormones.

- Our cycles are daily, monthly and yearly.

- Tracking your emotional and physical symptoms each day, week or month can give indicators to your hormone pattern.

- Your hormone pattern is unique to you and your life; there is not one type of woman. Being a woman is about what you identify as and your hormonal pattern is your own way of tuning in to yourself.

- Knowing your pattern can give you superpowers ...

MONTHLY CYCLE: MENSTRUATION

Have you ever wondered why you feel laser-focused one day and foggy the next? Why some days you feel hungry for just about anything, and others you just don't have as much appetite? Have you had those points in the month where your libido is off the charts? This cycling through moods and symptoms is all part of the four phases of our hormonal month. Our monthly cycles are so much more than just when you bleed, ovulate or feel the onset of premenstrual syndrome (PMS).

You might already know the three hero hormones associated with your monthly cycle. These are: estrogen, progesterone and testosterone. However, although crucial, these three are far from the full story of how your body shifts day to day, week to week and month to month. There are in fact seven key hormones that are triggered through the four phases of your cycle. They each rise, fall and, like a relay race, pass the baton from one phase to the other, taking you from a monthly bleed to ovulation and back to bleed. In effect your body has a new start almost every month, allowing it to recharge and realign. Here's a quick rundown on the seven hormones at play:

Estrogen. For women, this is the superhero hormone. It triggers rising moods from bleed phase through to ovulation. As well as elevated moods, estrogen can give you drive and motivation in the first half of the monthly cycle. This is the hormone that signals to the outside world that your body is ready to mate. Studies have shown that rising and higher estrogen can create a 'glow' and an increase in perceived attractiveness.

This includes more even and illuminated skin tone and thicker or fuller hair. This is our superpower seduction hormone.

It is also the hormone that is responsible for fat distribution across the body: think boobs and curves around the hips and bum, which are indicative of female form. Often you can tell if a woman has a higher estrogen balance, based on her shape and curves.

Progesterone. This is the counterbalance to estrogen. It is the calming hormone, and it kicks in post-ovulation, as your estrogen levels dip. It works in harmony with estrogen to provide balance to estrogen's hyper-energising qualities. This sedating effect is brought on to help slow your body and mind down, as your body works harder internally to cycle through old hormones and prepare for a bleed and the beginning of a new cycle. This means that it can bring lower energy and more introverted behaviours. High levels of progesterone can make women more sensitive to pain in the second half of the cycle. This is a good sign from nature that we are supposed to be more calm and restful in our higher-progesterone phase.

Testosterone. Although women have lower levels of testosterone than men, this hormone also rises and falls throughout the monthly cycle and is often linked to an increased sex drive both around ovulation and just before the onset of the bleed phase. It is often signposted by an increase in the amount of cervical fluid, which can also be linked to higher libido or arousal at different points of the month.

Follicle-stimulating hormone (FSH). This is by name and by nature the stimulant hormone and supports the maturing of the egg follicles within the ovaries ahead of release.

Luteinising hormone (LH). The release of this hormone is the green light for the egg to be released around twenty-four to thirty-six hours after peak.

Gonadotropin-releasing hormone (GnRH). This is the hormone that signals between your brain and ovaries pre- and post-ovulation. It is the hormone go-between, your menstrual referee.

Sex hormone binding globulin (SHBG). This is the delivery driver of the hormones estrogen and testosterone. It sends them to target tissues in the body and influences their function.

But we live in a world where, a huge proportion of the time, we have sex for pleasure, not just for reproduction, and these hormones do more for us than simply make us fertile or not. These effects play out in boardrooms as much as they do in bedrooms, so it is important to understand how they can have an impact on our social or work milestones. The effects of these hormones on our fertility is something that we tend to be reasonably knowledgeable about. Those who are trying to get pregnant may be familiar with the concept of fertile windows and all the other terminology associated with the hormonal phases. But these hormones do more than that: they affect everything from when to ask for a pay rise, or when you might be more likely to snap at your boss, friends or family. When we tune in to our pattern, we can use this information to prepare and therefore optimise for our best and worst days. The secret is to attend to your body's rhythms and cycles.

As I will continue to reinforce, there is no one-size-fits-all: every woman has a unique hormonal blueprint of moods and symptoms. However, what is known from centuries of research is the average cycle length is twenty-eight days, and that there are four phases, from bleed to ovulation and back.

WHAT IS TRACKING?

Have you ever had moments in the month where you feel supercharged and powerful, but the very next day feel like a different person with no special powers at all? Emotions and physical responses connected to hormones that rise and fall in four phases each month can be tracked, and therefore understood and even forecast like the weather. This emotional weather forecast can be used for planning your weeks and months ahead. You can focus on elevating the good days and preparing for the bad ones – knowing when to pack an emotional umbrella as there is a storm brewing. As Dr Minisha Sood, a leading endocrinologist in New York, puts it: 'The four hormonal phases of a woman's cycle can be tracked to gauge a baseline for physical and mental health. Using this data can help any physician get a quicker and more effective diagnosis if there is something wrong. This information helps focus what to test for in cases of hormonal imbalance. These phases are the foundations for women's health and well-being.'

Tracking cycles has been done for centuries in a pen-and-paper format. You simply mark out the four phases of your hormone cycle on a paper calendar, based on the last bleed you had. You then record what changes you feel each day and week within the four phases across your cycle. Within the Moody community the most common moods and symptoms women log are energy, anxiety, fatigue, motivation, bloating, headaches, breast tenderness, cramps, thrush, constipation and diarrhoea. These are all directly affected by hormones and finding patterns in their frequency across your cycle is a key indicator for what food and exercise could help at different stages either to reduce their impact or even completely remove them. The first step is you identifying what and when your changes happen.

What struck me about this age-old method was how ineffi-cient it is, with women having to continually go back through journals and look for patterns. It is like asking an author in the twenty-first century to write a book with a pen, paper and no automatic spell or grammar check: it can be done, but it would take twice the time. In the early stages of research and devel-opment for Moody, it became more and more clear to me that not only was tracking key, but that there were ways of making it easier through technology. Technology simply streamlines the process and makes the patterns clearer and the answers more accurate.

Tracking is an extremely valuable tool; it extends far beyond being familiar with when your bleed is likely to start or what your mood might be like. Cheyenne Morgan is a registered nurse, one whose own experience of cysts and endometriosis has led her to have a passion for gynaecology, and for communicating with women about their gynaecological health. Seeing how hard these conversations can be, she has started an Instagram account called @letstalkgynae to make them less scary. She describes what tracking her cycles has done for her: 'I see it as a big asset. Life-changing. For me, being able to track my moods, I feel like I'm booking an appointment with myself, for me to evaluate what is happening within my body. To be able to tune in with my body a lot more and understand, in terms of my emotions on that day, where I can make changes in the foods I am eating, in the kinds of activities I can do.'

What is true of all people is our need to better understand ourselves, and tracking technology and data can offer a mirror back to us, if we design the tech in a safe and inclusive way. It has been a powerful and eye-opening experience to see just how beneficial tracking can be to all women in our community.

Hannah Winkler is a trans woman who has been using tracking to understand the patterns of her body during her transition. 'When you begin transition you start taking a lot of hormones and at the outset, for the first three months, it is hugely turbulent ... Tracking these moods and symptom changes brought on by taking estrogen helped me rediscover my new physical and mental self. Tracking also helped me identify that I was having a cycle, as I would have monthly cramps from the estrogen I started on. I was able to see there was a pattern to these cramps that would repeat each month, giving a sense of my 28-day cycle. This has helped me gauge the effectiveness and range of the levels of estrogen I should be taking, and helped my doctors see both mentally and physically how the dose and the levels are affecting me.'

THE PHASES OF THE MONTH

Below, I outline in more detail the specific hormone changes at each phase and average length for each. The more you understand these changes, the more prepared you can be each month for good and bad days.

The key to tracking is to establish a baseline of your cycle phases and the moods and symptoms that tend to affect you most. What you're looking for are the anomalies or consistencies across your month. When you track your baselines, you can then better identify if things start to change.

The four phases of your cycle each have a set of emotional signals. When you tune in to them, it gives context to patterns in your body confidence, shape, facial expression, cravings and moods. For instance, it's common to find yourself reaching for pizza or chocolate when you're a couple of days away from the start of your period. This craving for comfort foods during this phase is linked to progesterone, which is known to interfere with serotonin production at this time. This may account for your body's desire for serotonin-boosting carbohydrate foods like bread or chocolate. It's not your lack of control or bad behaviour, it is your hormones screaming out for those treats, to make the final phase of your cycle a little more bearable. It's actually sometimes good for you to listen to what your body needs and evoke some happy hormones to help level out the PMS.

First, I will give a rundown of the four phases as the foundation for understanding these monthly seasons. The phases are pretty clearly defined when you begin to tune in to them. The beginning or end of each can vary from woman to woman and relies on you having a period each month. Even if you are not having a period, tracking your moods and symptoms across a monthly cycle can help give signs as to what might be blocking your regular bleed.

No matter at what point in the cycle you start tracking your phases, you will begin to build a fuller picture of your overall hormonal health and well-being the longer you do it. Looking from an emotional and physical well-being perspective, I have always found that starting with the bleed phase feels the most logical place to begin. Day one of bleeding is a clear trigger moment each month: the point when the sedating hormone progesterone begins to slow and the elevating estrogen begins to rise again.

PHASE ONE: BLEED

The bleed is triggered by the corpus luteum (which forms from the ovum follicle in the fourth phase of the cycle) magically disappearing, as it is reabsorbed by the body, signalling to your uterus that the lining is no longer needed and should be shed. It is this lining that constitutes the blood, but alongside the blood are other cells and no-longer-needed materials produced as part of the fertility process. This blood is often heavier and crimson at the start, while getting thicker and darker (sometimes even brown) by the end.

I am not going to say bleeding for several days a month is some kind of bliss. For many women this can be a debilitating time, as our brains become far more susceptible to pain in general. Pain during this time can be completely unbearable for some women and the experience of heavy bleed, or crippling cramps, can even render some people incapable of leaving the house. This can be linked to conditions such as endometriosis, but it is only in the last five years that such conditions have taken centre stage in the discussion on the secret suffering of so many women, who have mostly been told it's just 'that time of the month' and to 'get on with it'. This dismissal of physical pain over time has longer-lasting mental implications as well. Pain is of course relative to the individual, but it's an important symptom to track and understand, as it should be the first sign that something isn't right. Pain is your brain and body's best neurological flare signal to stop and assess the problem and seek help if necessary.

Bleeding for several days is sometimes also messy and, although just part of the process, not always clockwork in its timing. The key is not getting caught out by this phase. There are signs you can look for, based on your own hormonal patterns, and it is always worth logging and even retrospectively recording specific changes in moods and symptoms in the two days leading up to your bleed. This can help in giving you a much better idea of when you might be early or late. For me there are two days before my bleed when, along with low moods and bloating, there are more pronounced patterns in my appetite and libido. At this time, no matter how much I eat, my appetite is never satisfied and I also crave sex and pleasure like no other time in my month. Along with millions of women, I have been caught out so many times by coming on my bleed early or late and finding myself wearing cream trousers that I bleed straight through in a full Lady Macbeth horror scene. Getting caught out sucks, but by forecasting each phase you can be ahead of your bleed. Now, I always know when it's OK to wear cream or stay safe with black jeans.

Wherever we stand on the scale of painful periods, this bleed phase forces most of us to slow down, for a few days at least. Have you ever noticed how at this time your body encourages you to be kinder, gentler and not push yourself too hard? This is a time when often we crave sweet treats as they give us extra energy and can support our lower iron levels during bleed. The best antidote is low-sugar and good-quality dark chocolate, as high cocoa levels are a great natural remedy to increase iron and zinc. Have you ever felt horny on your period? It's common around this time, as your brain is more tuned in to crave intimacy. And when you have sex, climax can often be even more sensual, as your brain is more emotionally sensitive.

Like any cleanse, this bleed phase is part of the body's way of regenerating and recharging, making room for a fresh start. We are full of renewable energy and this phase is a flush-out of old proteins, cells and sex hormones, ready for your body to begin producing more-energising hormones including estrogen.

While your body is focused on shedding, your brain is tuning up ahead of a new cycle. Your brain has two sides, with the left controlling our logic and being linked to the more mathematical or analytical activities, and the right side being our imagination, intuition and creativity. Both the left and right sides of your brain are hypersensitive during the bleed phase, which evokes higher emotional intuition – your magical sixth sense – along with more analytical thinking. This phase is the time of the month when we should be planning and thinking about decisions ahead, utilising this practical intuition to help guide our ambitions and progress. Take advantage here of your improved memory and cognitive activity to plan for a productive month. The bleed brings with it a natural focus that can ensure your objectives are balanced between blue-sky ambition and realistic expectations.

Interestingly, I have also found that this is the moment when most women who are using the Moody app tend to log more memos and journal entries. Across the global Moody community there is a 37 per cent increase in daily and weekly engagement. It's a time women want to plan and take stock. It is as though your conscious and subconscious are aligning, while your body is more vulnerable due to your bleed.

PHASE TWO: RISE (FOLLICULAR PHASE)

This is your power-up phase and it lasts seven to ten days. From a biological standpoint, what is happening is that your pituitary gland releases FSH and signals to your ovaries to prepare for releasing an egg.

What happens internally and externally during this phase is that you start to feel (and even look) your best self as your estrogen, LH and GnRH climb towards ovulation. Your body even smells different – in primal terms, once upon a time it meant you were looking to attract a mate. In modern terms it means it's the time of the month where you can take on the world. Your skin glows, hair flows and you tend to be at your most outspoken, ambitious and driven. We tend to feel less vulnerable and our extrovert qualities can be more visible.

This is the time to engage in the harder, more complex tasks for your month. It's a great time to negotiate with others – to ask for a pay rise or push back in some of those harder personal interactions with partners, friends or family: you will be more able to navigate and multitask. You have more stamina for mental and physical activities, and to carry emotional burdens.

Your pain threshold is higher during this phase, which means harder exercise is more manageable. You are more likely to hit a personal best and be able to go further in any exercise you choose. There is no magic hormone that helps motivate you into the gym, but there are points in the month when the experience of exercise will be less challenging, and the endorphins and happy hormones released afterwards will take you to new dizzy heights. This is also the time of the month that I engage in the barbaric act of getting a bikini wax. It is one habit I wish I could give up, but until I recondition myself to love my pubes, I at least choose to do it during the hormonal phase when my brain and body will deal best with the pain of having hair ripped from every follicle.

The occasional downside to these elevated hormonal highs can be a slight over-increase in energy and more disrupted nights of sleep. Your brain is working overtime and the heightened focus and memory can in some women encourage a tendency to overthink and overanalyse.

PHASE THREE: SHIFT (OVULATION INTO LUTEAL)

This is the point in your cycle where your LH and, crucially, your estrogen peak. Once these elevation hormones have risen to their ultimate level, a sharp injection of FH into the cycle prompts the release of the egg. This peak point is when your mind and body seem to align, and you feel superhuman, as you are at your most hormonally charged. This is also the two- to four-day window where you are most likely to get pregnant – obviously an extremely important moment for those trying to conceive, but also significant in all the other periods of your life where you are just trying to be the best version of yourself.

This is a moment when your body can give you an extra injection of hormonal superpowers. In theory it is when you should feel your best, most resilient and certainly most strong-willed. However, it is worth being aware that if this is a time when you find yourself feeling more overwhelmed or anxious, it can indicate an estrogen imbalance and is worth tracking so that you can choose certain foods or supplements to support the peak in estrogen.

There is a burst of testosterone at this point, which can often evoke a higher sex drive. The peak window around ovulation is also a time when some women can experience increased appetite – your body is working harder to achieve more functions, meaning we can crave more energy to support this burn.

You may also notice your discharge increase during this window, becoming clearer and more lubricating and tending to look a bit more like egg white. This is simply your body's way of making having sex a lot easier. It's a good symptom indicator for when your ovulation window is, though it can of course vary from person to person. Following this moment, you can often see less discharge, which may be thinner and more opaque.

Next, your LH and estrogen dip, while progesterone takes up the hormonal baton and begins to rise for the final phase of the month.

Your body literally shifts from one set of hormones into a new set that prepares the way for cleansing and resetting. The recovery phase begins. As with any race, you need equal measures of rest and recovery to keep momentum: this is where progesterone kicks in. In this phase, as progesterone rises, you can feel your moods and sense of self become more introverted and women often experience a slower pace and tempo in their communication and drive.

This shift phase is a great time to simply take things a little easier: you will be achieving the same amount as you usually do, but be aware of your mind and body's capacity at this stage. Remind yourself that your brain isn't as lateral in its thinking and may be more tuned in to single-focus tasks. It's the time of the month I always lose my keys! Memory can often be less focused, so simply making sure you take more notes and give yourself reminders can be helpful.

PHASE FOUR: REFLECT (LUTEAL PHASE)

This can range from ten to fourteen days and is the phase that often evokes lower moods, higher anxiety and shorter patience or attention spans. It is the time when our bodies need as much self-care and support as possible. We stop producing LH and FSH and as our levels of progesterone get higher and higher, we begin to find a slower pace. We can become more irritable, and women often report having a very short fuse at this point. Although there is another burst of estrogen and testosterone just before the onset of the bleed, as a general rule during this phase your mind and body are resetting and should be resting.

This is also the phase where our pain thresholds become more sensitive. Our temperature rises, which, although it can make us more restless and vulnerable to external stimuli, can also mean much deeper and more intense sleep.

During this phase, where feelings of anxiety can increase and focus is harder to hold, it is a great practice to provide simple positive reminders to yourself each day. Firstly, the reminder that, much like the weather, this phase passes. Second, this is a moment that can be used to be more introspective and employ as much self-care as time will allow. It is when you should be trying to access happy hormones through all the activities that bring you most joy. For me this involves making plans with friends and laughing as much as possible, as well as taking long baths with books and listening to back catalogues of nostalgic nineties hip-hop.

PMS symptoms during this stage can vary from headaches, cramps and bloating to more extreme ones such as breast pain or tenderness, acute mood swings, cramp and aching across the body and limbs, headaches, fluid retention in ankles, neck and waist, insomnia, depression and anxiety. These may be an indicator of progesterone dominance or imbalance. What is key, when tracking these symptoms, is to record both when they start and when they end. This can help identify a pattern between more pronounced hormonal changes and triggers that could be affecting them. It is also useful to track what cravings you are experiencing or foods you are reaching for during this time, as it can help you to establish an effective support routine in your diet and exercise ahead of PMS kicking in each month.

Unlocking happy hormones at this phase is not just about helping with the lower moods, but is also important for pain relief. Although pushing yourself with exercise at this time can be fruitless in the quest for a personal best, it is still worth accessing the endorphins and pain-relief hormones that are administered after a workout. This exercise can be low impact, such as yoga, swimming or stretching, but any physical activity at this stage will help alleviate the higher levels of anxiety which can some-times make us feel more vulnerable and far more susceptible to criticism or likely to make impulsive decisions. This is the point when I have to lock my phone in a top drawer, so I don't text or contact anyone I shouldn't.

It is also the time when I try and avoid alcohol. Your body doesn't metabolise it as quickly, as your liver is working overtime clearing out old hormones accumulated during the cycle. This basically means you tend to get more drunk and due to the effects of high progesterone, you also can feel higher anxiety and lower mood, this paired with the known negative impact of alcohol and hangovers, means this phase can turbo not just how drunk you get, but more extreme downers and hangovers. The toxicity from the alcohol also stays in your body longer as due to all the other processes going on during this phase you don't have as much processing power from your liver. Adding excessive alcohol into your luteal phase is essentially creating an internal atomic bomb.

These four biological phases are a repeating cycle that lasts from puberty to menopause, but it is a cycle that can be affected by other hormonal cycles, including metabolism, sleep, stress and disrupters such as the contraceptive. Each phase comes with its own emotional landscape. Mapping out your own landscape through tracking will mean the more you connect with your baselines, the more you can be tuned in for when things slip off course. Thousands of women using and sharing stories via the Moody Month app and community have reinforced the defined nature of these phases.

The evidence is clear that the four phases are fundamental to hundreds of thousands of women. Although they may vary in length, moods and symptoms, there is no doubt that women have four seasons to their menstrual month. When these seasons are clear and follow a rhythm or sequence it's a good sign your body is happy and healthy.

DISRUPTIONS

Life is full of unexpected changes and new paths or choices. This means the rhythms and cycles of your body may shift in line with these disruptions. There are some common disrupters that we can monitor, such as diet, exercise and medications. Tracking will help us identify if they have had a negative impact on our body's natural superpowers.

One of the most obvious disrupters to our cycle are contraceptives which are specifically designed to act on our hormones. These monthly contraceptives are a billion-dollar pharmaceutical market. Although the hormonal pill alters the natural rhythm of your unique cycle, if you're taking the pill you still have phases; they will be different to our natural ovarian phases but they are still there. The differences will be chiefly noticeable in your bleed and rise phases. Specifically, women report less-elevated highs each month, as they are not having such a rise in estrogen, which is often the hormone that provides additional energy and prompts positive moods at these times. This is because the pill has synthetic hormones, which chemically trick your body into thinking it's pregnant. Some pills mean you still have a bleed each month, with others stopping the bleed process altogether. A bleed is helpful in that it denotes the beginning and end of your cycle, so it tends to make tracking your moods and symptoms each month easier. You can follow the pattern of your cycle in the same way from bleed and back, but you will not have the same defined ovarian phases or seasons. If you have no period at all, it's worth looking for signs such as breast tenderness, bloating and water retention, to assess when your body may be in its rest and recovery phase, and use this as your start and stop point each month. The reality of the pill, as with our natural hormones, is that everyone has slightly different patterns of moods and symptoms.

The main things to look out for, which can be signs that the pill you are taking isn't agreeing with you and your natural balance, are mood swings, weight gain or loss, hair growth or thinning on face, back or arms, and constipation. It is worth tracking any changes that happen when you go from a natural cycle to a synthetic cycle, but mostly so you can just support your body's needs each month. Emotional changes can be harder to notice, but tuning in and being mindful that these could happen will help in establishing a pill or contraceptive that works for you and your body.

There are a series of contraceptives on the market that have different impacts on your body's natural hormonal balance. These include:

Pills (or oral contraceptives). Brands vary and each will have a different level of estrogen and progesterone; this is why tracking any adverse moods or symptoms you experience can be an indicator of whether the higher levels of each are having an adverse effect on the natural balances of these two power hormones.

Patch. This contains estrogen and progesterone, but is administered by being placed on the skin each week. Your skin is an organ so can absorb certain levels of these hormones (a fact which also shows how anything you put on your skin can affect your hormonal cycles).

Ring. Another format to distribute estrogen and progesterone into the body. The ring is inserted inside the vagina and hormones are absorbed by the vaginal lining. The ring is replaced on a monthly basis.

Birth-control shot (Depo-Provera). The injection contains only progesterone, and is administered every twelve weeks by a doctor. Due to the levels of hormones given, it can take twelve months for your body to realign or shift back on to a natural hormone cycle once you stop receiving the shot.

Coil or intrauterine devices (IUDs). There are IUDs with and without hormones. In ones that release hormones, they can contain progesterone. IUDs are inserted into your uterus by a doctor and must be changed every three to ten years, depending on the type.

Implant. The implant contains progestin that is released through a thin rod embedded into your arm by a doctor. It lasts for up to three years.

There are over 200 possible brands and versions of the hormonal contraceptive, though it is important to understand that some pills effect some women in different ways. Adverse mood and symptom effects from hormonal contraceptive will depend on your body's natural baseline of estrogen and progesterone, as what the hormonal pill does is change this baseline. For instance, women who have underlying estrogen dominance can feel more balanced when on a progesterone pill. However, the pill can also mask symptoms of conditions like polycystic ovary syndrome (PCOS), which women sometimes only discover when they come off the contraceptive. Not all pills give people adverse effects, but they do change your hormonal make-up and balances.

There are also non-hormonal alternatives such as the non-hormonal coil and condoms. If condoms are your choice, it is worth making sure you choose one that suits your body. Dr Sarah Welsh, a gynaecologist, and Farah Kabir founded HANX, the first condom company designed by women for female pleasure. They point out: 'Condoms have been designed with men in mind for decades, all the way from the packaging, the messaging, the sizing and the materials. Yet these products go inside a woman's vagina. The industry has been designed around virility, male strength and size. No woman I have ever met wants to think about a Trojan horse while having sex! It has not been a woman's world! We think about our skincare and what we put on the outside, yet for some reason we have not been given options for what goes inside us, within the most intimate moments of life.' Their products keep in mind women's bodies, and their vaginal pH balance. Made from 100 per cent latex, they don't use chemicals commonly found in condoms such as anaesthetic (which help men last longer to enhance their sense of sexual success, but can also irritate a woman's vagina).

It seems crazy that we are only having these conversations and developments in 2020, yet these experiences and issues have been part of women's lives for centuries. Shardi Nahavandi founded Pexxi, a digital platform to enable women to be given the best-possible information and choice about contraceptives, based on their own hormonal profile. It arose out of her own experiences of medical problems not being diagnosed properly – even to the point of being misdiagnosed with bowel cancer. A particularly horrific encounter with a contraceptive pill that her body couldn't tolerate led her to realise that the whole current approach to this area needs to be updated: 'There are a lot of myths in the contraceptive space, and we want to help re-educate on how misinformation has led women to taking the wrong products.' She is working to revolutionise the industry out of its one-size-fits-all template. The pill is not simply a pregnancy-prevention device – there is, as she puts it, a 'massive therapeutic aspect', with contraception being prescribed for a range of different conditions. 'And yet every single study done on the effects of the contraceptive is based on how effective they are at preventing pregnancy. It is a commercial business model about preventing pregnancy, which in turn helps avoid the other financial costs to healthcare brought on by unplanned pregnancy. This business model does not have women's well-being at its core. We want to look at the contraceptive as one option, and at how your biological data can help match the right option for your emotional needs and physical body, not just as a prevention tool ... ' At the heart of it all lies education: 'If you decide to take the contraceptive, you should be informed about the potential impacts certain types may have. With hundreds of contraceptive brands, rather than being given what is cheapest, you should have the right to choose, based on knowledge of your hormones and your body.' As Shardi so strikingly says, 'Contraception is health; health is contraception': there should be no division between the two.

There is no conclusive science to say the hormonal contraceptive pill has any impact on your long-term health, but there is also no research that has been done to disprove this either. Adverse emotional and physical effects will depend in part on your genetic and hormonal make-up, which vary from woman to woman. Even though the science and research is so far behind and hasn't given us any specific conclusive information, what is known about our bodies is if we ignore or dismiss adverse emotional or physical effects from any external factor, it will lead to profound effects on mental and physical health. Which some women only realise when they subsequently come off their course. This is another reason why keeping track of how the hormones in your cycle change month to month can give you the best indicator as to whether a particular pill suits you.

I am a big believer in choice and everyone should have the right to take the pill if they choose. This product has had huge social benefits to women since its launch in the mid twentieth century. However, what troubles me in the twenty-first century is how little women are asked about their cycles, hormones, and health generally, before being given an incredibly powerful drug that can alter body shape, emotional well-being and even our sex drive. Choices are about being armed with the right knowledge, so you can make an informed decision for you and your body. As a nurse, Cheyenne Morgan agrees that it is important for women to access the right information about contraception: 'Talking to some women … when it comes to contraception, a lot of them don't know a lot. It's more Google-based information.' It's her mission to demystify this area.

And the good news is there are now resources out there to help you make this decision. Susan Masters is the national director of nursing for the Royal College of Nursing in the UK. She has worked in patient-facing clinical areas for many years and now advocates for nursing across the country. Nursing care is a critical part of sexual health and hormonal health diagnosis: so much is rooted in the conversations nurses can have with patients to identify patterns of symptoms. Susan underlines that when it comes to contraception, everybody has different needs, and that this has become a vital part of sexual healthcare. She has identified a huge change here from the days when a sixteen-year-old girl had to be accompanied by her mother to see the GP in order to access contraception, and when there may have been a generalised, blanket approach to types of contraception, based on broad-brush categories like age: 'What is different now is that there is a lot more individualised, personalised care. Nurses have a huge amount of education on how to communicate with patients about their options with contraceptives. We look at the individual's lifestyle, body and circumstances – are they in a relationship? Sexual health being run by nurses does give more opportunity to personalise care, and also to help pick up risks. This is true of all care: having conversations is key. It is all part of the understanding of how we build a bigger picture of the nuance of a patient's life.' Susan points to the fact that there are now sexual-health clinics run entirely by nurses in order to help women have these discussions. In other words – if you are looking into contraception, it is important to find the right route for you, and there are nurses who are trained to help.

The contraceptive pill has helped and supported many women in avoiding unplanned pregnancy. But there is a huge social issue as to why society has placed this hormonal responsibility solely on women. The male hormonal contraceptive is now coming to market (which, considering the female hormonal contraceptive launched last century, seems a little overdue), but the responsibility for women to 'protect' themselves remains a long-term social expectation, something that may not be solved simply by inventing a new pill for men.

DIFFICULTIES

Life isn't all good news and happy highs, which is why being moody once a month is actually a great way to indicate that your hormones are cycling through their seasons. However, there is a big difference between having four phases with some lower moods and uncomfortable patterns, to having a condition that is directly connected to the very hormones that cycle through each month. Such conditions can make that time of the month not just an inconvenience, but a debilitating and sometimes even life-threatening moment. They vary in severity, but they are sadly not uncommon.

One key indicator of premenstrual conditions are irregular cycles. PCOS affects 1 in 10 women and endometriosis affects almost 200 million women worldwide. They can have debilitating results, with symptoms brought on each month as the hormones cycle through your body. PCOS can also cause difficulty for women trying to conceive. Along with irregular periods, the indicators can also include other things connected to vast hormonal fluctuations, such as excessive hair growth on face, chest, back or buttocks, weight gain and oily skin or acne. PMDD is premenstrual dysphoric disorder. This clinical condition was only taken seriously and classified by the World Health Organization in 2019. The symptoms mirror those of chemical depression and include feelings of despair and hopelessness, intense anger and even violent behaviours, tension, anxiety, loss of or increased appetite (under- or overeating), lack of concentration and fatigue. It is a condition women have been living with undiagnosed for centuries and, although finally there is more public and medical awareness, they are still suffering in silence, as there is a dismissal of these extreme symptoms as being just part of your cycle. Knowing when in your cycle these pronounced symptoms reoccur can help get the quickest, most confident understanding of whether you're suffering from PMDD and how best to support it.

Living each month with the severe pain and emotional symptoms connected to these conditions can even impact women's physical ability to leave the house, having a knock-on effect on work and relationships. It can mean women who suffer from them slip into other forms of mental and physical illness, including depression and eating disorders. The recurrent nature of the symptoms can cause a build-up of pain, fear and stress, but the societal dismissal of the conditions as 'menstrual' means many women go years or even lifetimes without proper diagnosis. Women I have met while researching the menopause for Moody have shared harrowing tales of a life living in extreme pain and only being released from this monthly prison when they hit their fifties.

A lot of the female doctors we are working with at Moody are having to re-educate their male counterparts on the severity and importance of these conditions for women's physical and mental health. Saschan Fearon-Josephs founded the Womb Room web platform after being diagnosed with endometriosis and fibroids. Her journey is one of true bravery and resilience: she was met with dismissal and a dangerously late diagnosis for a condition that could have been fatal. As she puts it: 'Explaining symptoms to doctors is so hard and it often takes multiple visits for a doctor to see the patterns. What is often worse is feeling you're not being believed, both by doctors and by the people closest to you – family and friends.'

All of this is why tracking can be so valuable – it has the potential to give every woman a language through which she can talk about her body. Tracking the patterns of your symptoms and cycles can help fast-track and support the diagnosis of conditions such as PMDD, endometriosis and PCOS, ensuring you get more medical support. Clinicians work well with evidence so tracking can reduce ambiguity and provide solid information for them to support any diagnosis. Susan Masters underlines this point: 'PCOS, endometriosis and PMDD are becoming clearer with diagnosis, as nurses are unpicking the root through conversations about pain and patterns. Tracking does help with this, as it makes these conversations quicker and easier if a patient has more insight and information about their past and new symptoms. It makes the nursing conversation more of a team approach between patient and nurse.' She also points out how easy it is to let these symptoms pass unnoticed, which is why making tracking easier can be so important: 'As a woman, your hormones, your menstrual cycle ... you deprioritise it ... So I think that a quick digital platform can be really useful. Anything we can do to bring women forward to get symptoms looked at and assessed properly has to be of benefit.'

One particular area came up with several different people who I spoke to for this book: that of how we articulate pain. Cheyenne Morgan was especially interesting on this subject: 'Often we ask women to explain their pain on a scale of 1 to 10, and this is not serving women well in helping to identify someone's true pain. They can't explain it just as a scale. We need to dig deeper in understanding the type of pain, but also how these conditions are impacting their wider lives.' Having suffered from endometriosis herself, she feels the questions should be: 'How do your moods connect with your pain? Does it interfere with your relationship; does it interfere with your social life? How long has it been going for?' I liked her final point: 'Describing pain, both emotionally and physically, is something we should all learn in school. It would help us all communicate better both with our closest friends and family, as well as doctors and nurses when we need them.'

Saschan Fearon-Josephs echoed this view: 'I did a round table with the Royal College of GPs to try and understand how doctors are seeing and hearing patients – it means they don't get triaged in the right way. If you can't explain what your pain is, where it is, as you don't understand the basics of your physiology, then you can't possibly communicate this to the doctor. We often expect the doctor to be able to diagnose us from what we say, but nobody has given us the language to help us communicate effectively exactly what pain is "normal" versus "extreme".'

As she also points out, the danger is of 'having your symptoms dismissed as PMS or just something you should put up with'.

Tracking can help here, giving you a clear picture for yourself and medical professionals of what is really going on. Here are the symptoms to track if you think you are suffering from PCOS, endometriosis or PMDD:

PCOS: Irregular periods; no periods; acne or oily skin; excessive hair growth on face, chest and back; thinning or loss of hair on head.

Endometriosis: Pelvic pain and extreme cramps; a heavy bleed; bleeding between periods; bowel pain and cramping; pain during sex.

PMDD: Depression or feelings of hopelessness; intense anger and conflict with others; tension, anxiety and irritability; fatigue; changes in appetite.

If you are experiencing any of these symptoms in your cycle, then track, log and take this data to your doctor. It is powerful to go with specific dates and times for when these recurring cycles kick in. It is also empowering to know that these are important conditions, ones you should be addressing and getting support for.

Patterns and cycles are not just key to effective personal care and management, but also the information helps give power back to you and enables you to ask focused and useful questions. It's the advice that I will keep returning to: you are the best person to listen and understand your body – you're the one living inside it.

STRESS AND ITS IMPACT ON OUR CYCLES

Aside from endemic health issues such as these, another key reason for disrupted cycles can be stress. This often comes with a suite of other symptoms, so it is worth tracking what environments, people or foods trigger certain stress symptoms, as this means you can begin to map how stress is affecting your body physically and mentally before you hit burnout. Key signals and signs to track are extreme fatigue, weight gain and gut issues, including IBS symptoms such as gas, bloating, constipation or diarrhoea. Stress can also impact on appetite in some women, causing weight loss which can then in turn disrupt estrogen balance and reduce ovarian-cycle function. Often when your BMI drops below a healthy level, your cycles will also shift into a crisis mode. Women who suffer from anorexia nervosa often lose their monthly cycles and this can contribute to other cycle disruptions across sleep and metabolism long term.

These symptoms can be a sign of an overstimulation of your stress hormones and can put your liver under extra pressure, due to the added toxicity in your body. Your liver is the key organ for processing and cleansing your body of toxins and old hormones each day, week and month. When my own burnout hit and my stress hormones were off the charts, my liver became overwhelmed. My elevated cortisol was interfering with my thyroid function – the gland that regulates metabolism. There is a triangle inside all of us that connects our menstrual cycle with our stress responses. The three points of this triangle are glands: thyroid, adrenals and ovaries. They each secrete hormones that help inform each other and regulate various cycles in response to stress across the body. When one is overproducing or underproducing, it has a knock-on effect on the rest of the triangle.

For me, stress had an impact on my thyroid and ovaries. The obvious symptoms of periods stopping and my gaining weight were both due to disruption to my triangle. My body stopped processing fats and started storing them around my middle, bum and hips. I also started growing massive boobs: though at the time it felt novel, the overall weight gain was mentally debilitating and making me foggy and fatigued. Sadly, there was no magic pill to fix this. The remedy came from knowing

THE HORMONE TRIANGLE

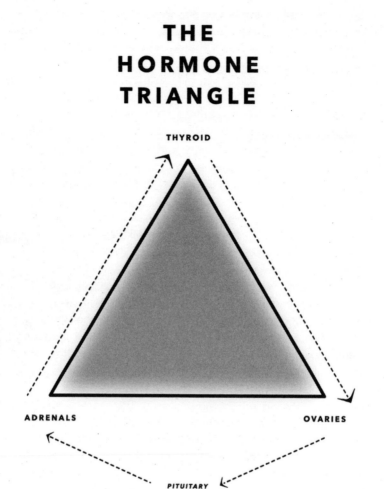

THYROID

ADRENALS

OVARIES

PITUITARY

the root causes of the problem and making a few simple lifestyle changes. I worked with my nutritionist, Lola, who helped repair my very sad and overworked liver, and the results were both physical – weight loss – and, more importantly, mental and emotional. In simple terms, I got back my mental clarity and happy hormones.

THE SKIN YOU'RE IN

The biggest organ we have is our skin. Although we often think of our skin as something to treat or maintain, we can also regard it as a map to our inner well-being. And we can use this map to assess our levels of stress. Balancing your hormones can make you glow: your skin is the mirror to your insides and when your hormonal system is in check, you look like the goddess you are. On the other hand, spots, dryness and irritation can come on overnight or within our 24-hour cycle. These effects can be triggered by stress and imbalances or fluctuations with hormones. Most women experience changes within their skin's clarity and glow across the four phases of their menstrual cycle each month. For some these changes are profound and lead to breakouts in the fourth luteal and bleed phase, as estrogen is low or begins to rise causing more anxiety and PMS symptoms. There is a direct link between emotional stress and skin breakouts and a great way to address this is to support your hormonal changes with fibre-rich vegetables and supplements such as chromium to maintain blood sugar levels.

Across your monthly ovarian cycle, there are shifts in your skin response based on your levels of estrogen, progesterone and testosterone. From bleed to ovulation, as estrogen rises, this stimulates collagen and hyaluronic acid production. This promotes efficient nutrient absorption and in turn gives skin a glow. In the second half of your cycle, rising progesterone can trigger oil glands in skin and pores can compress, which may cause oil build-up. This, combined with testosterone rising, which prompts oil glands to secrete more sebum, can then cause hormonal spots and breakouts. If you are experiencing breakouts throughout the month, in particular around the chin, mouth and lower cheeks, it can be a sign of hormonal imbalance. This is your skin's way of sending out a flare, alerting you that you need to focus on helping it internally.

The contraceptive pill is often prescribed to girls as young as eleven years old, as a way of medicating hormonal acne during puberty. Although this can often have fantastic and fast results, hormonal therapy does not address the root cause of the acne and has meant many women suffering from syndromes such as PCOS have gone undiagnosed, due to the contraceptive masking many of the other symptoms brought on by the hormones.

What is interesting to look for when you get spots is where on your face they appear; this is a key sign of which glands and where your body needs some extra support. If spots appear around your chin and mouth during the reflect (luteal) phase of your cycle, this can be due to lower estrogen and can sometimes be addressed by omega 3 fatty acids from plant or fish sources (these can also be taken as supplements). Omega 3 will help improve hormone communication and reduce inflammation. One of the best and simplest remedies for monthly flare-ups is increasing your hydration. Drinking a minimum of two litres of water per day for the final phase of your cycle can keep your skin and body regulated. The beauty of reading signs like skin flare-ups as a way of mapping your internal system is that you have the power to address them with food and certain types of exercise, supplementation, and understanding the gut–liver axis and detoxification.

HOW TO KEEP BALANCE

Once you have identified the phases of your own cycle, and how specifically it tends to affect you, you can move on to addressing any imbalances. What I can share here are some of the top tips my nutritionist Lola Ross shared with me and I hope you share with other women. Lola has not only been a major factor in my own recovery, but is also a huge part of Moody's mission to make health and wellness information more accessible and democratic for women. She has shown me how diet and supplements can be used to enhance my body's natural processes. These are not 'bio hacks', but good routines and rituals to unlock extra levels of energy, metabolism, productivity, motivation, and even sex drive. Supercharging your cycle is very simple. It doesn't involve changing your diet, simply rotating or adding certain foods and supplements to your existing lifestyle.

There are specific food groups and natural supplements that support women's internal levels of estrogen, progesterone and testosterone. In turn, these feed your pattern and allow the natural cycles inside you to work at their best, including metabolism and sleep. It's just about knowing what phase you're in and the food that's best for achieving this balance.

PHASE ONE: BLEED

Exercise. This is a restful-tempo phase, so for exercise it makes sense to focus on low-impact activities such as yoga or long walks. Although I love to run, I try and do slower, shorter and more mindful running during this phase. No goals, just headspace.

Foods. According to Lola, foods that are good for your bleed phase will be about replenishing nutrients that your body requires for the new cycle. Proteins provide amino acids – the raw materials needed to rebuild your endometrium towards the end of this phase. Include plant proteins like tofu, quinoa, legumes, a scoop of algae like spirulina, and animal foods like eggs and fish.

Reducing inflammation in this phase is also often important for many women as it is responsible for cycle-associated pain such as cramps and migraine. Anti-inflammatory foods such as oily fish, plant oils, walnuts and avocados can be beneficial.

The mineral magnesium has a key role in nerve transmission and is known to act as a natural painkiller and mood stabiliser which can help to balance out any lingering PMS aches or anxiety. So magnesium-rich foods are good – try nuts and seeds, dark leafy greens (chard, watercress, spinach, baby kale) and wholegrains. Iron-rich foods such as meat, fish and eggs or plant foods like lentils and dried apricots can support iron-loss in those with heavier bleeds or flow-fatigue.

Supplements. There are some dietary supplements that can be really beneficial during this phase. If you suffer from cramps or menstrual pain, take a magnesium supplement or try a magnesium bath soak. EPA/DHA omega 3 fatty acid supplements and pyroxidine (B6) can support hormone regulation throughout the cycle and may reduce very intense flows and large clots. Iron supplements can be essential support for some women as iron deficiency is often seen in heavy flows, but due to toxicity risk you should check with a practitioner before starting any course of iron. Turmeric capsules have anti-inflammatory effects and have been shown to be effective in soothing menstrual pain.

PHASE TWO: RISE

Exercise. This is when your energy begins to rise. Pushing yourself harder and doing more high-impact exercise like running or weights will be a little easier, as your rising estrogen helps with your pain threshold; also your body will get a mega-boost and elevation from the additional endorphins. This is how to turn your power-phase hormones into superpowers.

Foods. This phase requires a balance of nourishing nutrients from a range of wholefoods in order to support energy and ovulation, while taking care not to overstimulate an estrogen-revved-up mind with sugars, caffeine and alcohol. However, so long as you drink sensibly, you may find that your body handles alcohol a bit better than during other phases in the cycle. Fibrous vegetables and fruits are always a good idea, as they keep blood sugar stable to help with energy, mood and metabolism, and fibre helps remove estrogen. Smoothies, soups and roasted vegetables are a few ways to get your daily fibre, vitamins and minerals.

If you find yourself a bit hyper during this phase, or overexcited about something in your professional life to the extent that it is keeping you up at night, keep calming food in mind. Try not to eat late as it stimulates your adrenaline, but if you do need a snack stick to tryptophan-containing foods. Tryptophan stimulates melatonin – your sleep hormone – and can help you feel more sedated. Try a warming turmeric oatmilk or half a banana in the evening.

Supplements. If your intuition is hinting that you might be a bit nutrient-depleted, or if you are limited on time, try a complete powdered greens supplement which is an easy way to provide you with a range of vitamins, minerals, amino acids and essential fats. Powdered form, or fresh green veggies, can support your follicular phase, delivering vital nutrients like vitamin Bs and potent cell-protecting antioxidants, and will help to keep your extra energy well supported.

For calming a busy mind in the evening, try the amino acid L-theanine, or consider magnesium for soothing the nervous system. Lavender or valerian infusions make great relaxants.

PHASE THREE: SHIFT

Exercise. You may have a surge of energy around ovulation, but this is generally a point where your tempo shifts from higher to lower. Don't be disappointed if it's harder to motivate yourself in exercise or make healthy choices in food.

Foods. Your hormonal shifts here can cause energy dips leading to cravings for instant sources of energy from sugars and fats. A spoonful of nourishing almond butter can be super-useful for stabilising your blood sugar and satisfying your taste buds so that you don't go cake crazy. Keeping blood sugar balanced is important as hormonal shifts in this phase can trigger anxiety. Try to eat foods low in refined sugars and high in fibre – think root vegetables like parsnips and sweet potatoes, low-sugar fruits like berries and fibre-packed wholegrains like oats.

Avoiding alcohol, chocolate, coffee and other caffeine foods is essential for those who suffer from intense mood issues and breast tenderness during the luteal phase.

Supplements. The high levels of circulating hormones at this time mean that the liver has more work to do removing used ones and keeping hormone levels in balance, which in turn keeps the cycle healthy. So a bit of additional liver support is a good thing now. Indole-3-carbinole are compounds found in brassica vegetables (e.g. broccoli, cauliflower, kale, sprouts) that have the capacity to enhance liver detoxification. Taking a daily prebiotic fibre formulation to support your gut bacteria is another useful tool. If you suffer from menstrual migraines, this is the time to introduce daily magnesium either orally or transdermally, until your bleed phase.

PHASE FOUR: REFLECT

Exercise. This is your luteal phase. Your energy will be on the decline and this is often when PMS symptoms occur. Opt for lighter exercise on a daily basis. A fifteen-minute morning routine at home will give you the DOSE and happy hormones you need to prepare you for your day ahead. This is another point when cravings kick in, so understanding how you can support blood sugar balance is also key.

Foods. Lowering hormone levels can have an impact on serotonin activity – the neurotransmitter that is involved in regulating our mood and happy feelings. Serotonin is largely produced in the gut by our bacterial ecosystem, so feeding our gut bacteria with prebiotic fibre-rich foods such as onions, bananas, beansprouts, leeks and most other plant foods, or even a couple of squares of high-cocoa-content chocolate, can in turn boost serotonin.

Supplements. Chromium has an influence on how our blood sugar stays balanced, which can have a positive effect on cravings, mood and anxiety at this time. A chromium supplement – while not a magic pill – might be helpful as part of a blood-sugar-balancing food plan, and cinnamon has some similar effects. Cinnamon is also a great alternative to adding sugar to cooking.

Evening primrose oil is a rich source of gamma-lineoic-acid (GLA) and can be useful in reducing the severity of PMS. Vitamin E supplementation may help if you suffer from breast tenderness, migraine and other cycle-associated pain.

I would never advise living religiously by these foods in each phase; it would take all the fun out of life. However, it is helpful to know at the beginning of each phase what your body needs in nutrients and how you can supplement or be more aware of what will work best and when.

WORKING WITH YOUR BODY

When it comes to exercise and looking after yourself, hormones are the great unremarked element. What has been missed from the huge cultural rise in wellness is providing routines that work for women's bodies and their monthly hormonal rhythm instead of catering for the male version, which operates on a 24-hour cycle. We have been conditioned by media and health trends to hit the gym every day and drink green juice like it's water, but for the way in which women's bodies work hormonally, this is not optimal – it's setting us up to fail. The fact is women's bodies process fats, sugars and energy on a phase-specific basis, meaning the phases can act as a guide to optimise the effect of healthy choices.

Adopting a routine that isn't designed for your body simply means you're more likely to fail at keeping it up long term. That will probably make you feel terrible, rather than unlock your happy-hormone power pack. Shara Tochia of DOSE echoes this point, reinforcing the concept that the aim of exercise and wellness should be to boost your levels of DOSE, not to put yourself under even more pressure: 'Even now I have to catch myself, as I work so hard, and then remember this can some-times make my body more stressed in the long term rather than happy and healthy. My routines for unlocking DOSE do change, dependent on if I am under work stress or feeling more relaxed. When I am stressed I avoid hard-core exercise, as although I am quite addicted to the endorphins from a spin class or HIIT class, I know the adrenaline, on top of my stress from work, will compound and mean I crash in energy later that day. Hard-core exercise is a tool for me to use and unlock endorphins when I am already in a happy place, and it supercharges that happiness, but it has the opposite effect when I am stressed and this is common for most women. It's about listening to the pace of your body and doing what workouts work for your emotional and physical state of mind.'

Nature designed women's bodies with phases that include rest and recovery. Playing to your hormonal strengths means your body will respond better both in energy production and metab-olism. When you're bleeding you should eat chocolate and put your feet up. Mother Nature wants you to take time out; it's part of being the half of a human species that can house human life. Women think laterally with multiple cognitive functions at any one time and have emotional intuition as a sixth sense, height-ened when estrogen is rising. All superheroes need to have a holiday to recharge their batteries and women's recharge is on a bimonthly basis.

As with weather cycles, you can forecast but never control. Your cycle can be unpredictable. Your body and life are being affected by so many more things than what you eat or do in the gym. We work, have relationships, friends, family and sex lives, all of which have important roles to play in our bodies' hormonal development and balances. Later, I'll talk about all the other parts of our life, and the hormonal patterns and clues associated with them.

We have been conditioned to think that running like clockwork is crucial, but this is almost completely counterproductive to the very nature of how our bodies change and respond to the world around us. I would argue humans are far more irregular than regular. Therefore, when thinking about the four phases of your cycle, it should not be to benchmark if you are 'normal', it should simply be used as a navigational tool for helping you to organise or understand when and how your own pattern works. What are your phases and what are the common moods and symptoms associated with these phases? Logging these, and building a picture over time, can help you to see when and where the commonalities and repetitions of moods and symptoms are occurring. What you are building up is a map of your own hormonal landscape.

TAKEAWAYS

- There are seven hormones involved in a menstrual cycle: estrogen, testosterone, progesterone, follicle-stimulating hormone (FSH), luteinising hormone (LH), gonadotropin-releasing hormone (GnRH) and sex hormone binding globulin (SHBG).

- There are four phases to menstrual cycles. These can cause fluctuations in emotional and physical patterns such as energy, motivation, memory, appetite and sex drive.

- The liver is a key organ in the body's ability to process through hormones from menstrual cycle. This means we can be more sensitive to alcohol at certain times of the month.

- Contraceptives are an important part of female sexual empowerment, but we should know more about how hormonal contraceptives affect our individual hormone patterns, to assess the right fit.

- Even when you don't menstruate, you still have hormone cycles. These cycles will change your moods and symptoms across your month.

DAILY RHYTHMS: METABOLISM

Have you ever felt the flow of happiness? 'Flow' as an emotional mindset is a term coined by the positive psychologist Mihaly Csikszentmihalyi. He draws on ancient philosophy and modern psychological studies both to categorise a state of mind that achieves human happiness, and to provide techniques that will allow you to reach that utopia. He argues that we're often overwhelmed with information and that we spend too much time comparing ourselves to others. In this era of social media, this natural tendency to compare ourselves becomes out of control and potentially toxic.

Getting into your flow state is arguably as much hormonal as it is psychological. When I read Csikszentmihalyi, what I see are tactics for accessing the happy hormones, DOSE, through the everyday pursuits that make us calmer or more focused. As we have seen, DOSE are those four happy hormones: dopamine, oxytocin, serotonin and endorphins. When we are in our flow states, the steady satisfaction of an activity that we are engaged with gives a consistent flow of DOSE and therefore our brains and bodies operate in a more focused and optimal way.

Flow states are key to finding focus and productivity, but they also offer long-term health benefits that can lead to living a longer life. Happy hormones keep your body's overall functions and everyday cycles – like sleeping and digesting food – working in sync. As humans, we operate on a circadian rhythm – a 24-hour cycle that directs our intersected sleep and metabolism. We are diurnal mammals, which means we are meant to operate around daylight hours. Although, for women, there are variable hormone fluctuations throughout your menstrual cycle that mean your sleep and metabolic cycles work at different rates at different times, the body still works towards having a day and night. Your hormones, such as the sleep-inducing melatonin that rises at dusk and the energising cortisol that rises at dawn, are key indicators of this daylight system. When humans operate around night hours this can have detrimental effects on hormone activity, disrupting melatonin or cortisol which can impact mental health or eating habits, as both these hormones are affected by light levels and triggered by the sun and moon cycles. Studies have shown that night-shift workers, including doctors and nurses, have increased risks of metabolic issues and weight gain, leading to higher risk of chronic illness such as obesity, heart disease and even some forms of cancer. It has also been shown that there is a huge increase in mental health conditions during winter months, markedly so in countries such as Norway where winter seasons have limited daylight hours. These seasonal affective disorders contribute to higher suicide rates, showing the power of sunlight on our minds and bodies. Having daily rhythms and routines that sync with our age-old circadian rhythms is essential for everyday and long-term health.

A DAY IN THE LIFE OF YOUR HORMONES

Have you ever had a bad night's sleep and then were unable to focus the next day? Felt 'hangry' when you get hungry? This is because when your body isn't able to tap into its natural state of recharge during sleep, or when it can't draw energy from food when it needs it, the response is mood swings, irritable behaviour and anxiety.

Your metabolism is the energy source cycle, where your body breaks down the food you eat and turns it into fuel. Your sleep cycle is where your body is resetting and recharging your glands and organs ready for a new day of energy, so the cycle works at different tempos during waking hours and sleep.

Hormones released each day and night tell your body what to do when, and how fast to do it. Cortisol begins its ascent in the early hours of the morning and then falls in the first hours of waking. Dawn is where our bodies begin a new waking cycle. Our metabolism begins to shift from a sleep state to a wake state as soon as we rise from bed. It is affected by when we last ate, as food that hasn't been digested from the previous day is processed; this is why ensuring you avoid eating at least two hours before sleep can help keep your metabolic system working effectively in the morning. By midday our bodies are in their peak daytime hormone cycle; this is when your body should have its optimal energy levels and efficient digestive function.

As we move through the day, the hormones connected to our appetite begin to play a part. These are:

Ghrelin. This is your hunger hormone and is produced to signal from your gut to your brain that you need more fuel from food.

Leptin. This is the hormone that signals you are full and suppresses your appetite.

Insulin. This is the hormone that removes glucose from your blood by allowing it to enter into cells to be utilised as energy.

The time of day you choose to eat is almost as important as the food you choose in regulating and optimising these rhythms. Intermittent fasting is effective as a way of losing weight, as you are giving your body natural breathing time between digestion cycles, to reset the important hormones involved; this includes regulating insulin and glucose levels, which go up or down as soon as we eat any food, no matter how healthy. Hormone health, as with most things in life, is not just about the choices we make, but also about timing. What is key about the end of the day is our cortisol begins to drop, GABA increases and melatonin is triggered before sleep. If you eat foods such as sugar at this time, it can limit the ability for your body to naturally produce these chemicals, which means it's harder to move from an active energy state into a resting sleep state.

The technical term for the internal optimal-rhythm regulation that keeps us alive is homeostasis. What happens when we're experiencing good sleep cycles and metabolising nutritious wholefoods each day is that we become synchronised with our natural rhythms. We don't get the negative and often distracting symptoms associated with disrupted cycles. The beauty of understanding how your daily rhythms work is that you become more aware of how to support yourself during the inevitable moments in which stress hormones impact your sleep and metabolism.

UNDERSTANDING THE GLANDS

We have seen that glands are involved in distributing your hormones around your body on a daily and nightly basis. Human understanding of the power of these glands dates right back to the first millennium AD and the Indian practice of mapping chakra energy points. Chakra points are positioned in exactly the same areas as our hormone glands. Knowing how 'energy', or hormones, flows through your body has been a cornerstone for human health optimisation since the year dot, yet we're now in the twenty-first century and it's still not common knowledge how these glands hold so much power in helping you to be happy and healthy.

Your glands are positioned throughout your body, from your brain to your ovaries. Glands are production and secretion tissue releasing their specific hormones into the system, which then carries them to their target organs to enhance or activate a particular function. They are your hormone control centres and determine what your body needs in order to survive.

MAJOR
ENDOCRINE SYSTEM

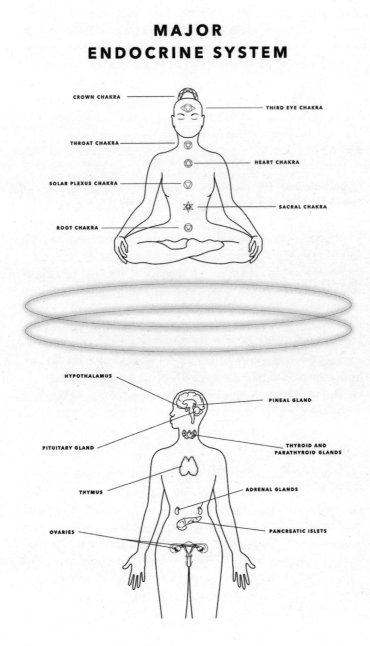

In order from top to bottom, here's a rundown of where each of these glands are and what their primary function is. I always think about these glands as our central core processors. If your body and brain are a supercomputer, then these components on your memory board or drive will determine how fast and effectively your system works. Each element has a core function, but they also only work effectively as a harmony of components. The negative moods and symptoms that can occur from hormonal unrest are due to one of these key pieces being out of sync. They rhythmically release their chemical compounds.

Hypothalamus. This is your central communication centre. Although the hypothalamus is a small gland, it is mighty. It functions as part of both the nervous and endocrine system so is considered a neuroendocrine organ, which means it not only connects the flow of hormones across our body but also how they make us feel and think. It is a major control across almost every bodily system, so it lies at the heart of your internal communication network. This gland responds to electrical signals from the nervous system by sending out chemical messages in the bloodstream which in turn stimulate the other glands. It controls the involuntary nervous system, regulating everything from digestion to blood pressure and even the rate of our heartbeat. It is part of the emotional region of the brain and influences our perception of fear, pleasure and pain, sweating and blushing; and drives biological rhythms like our sex drive and the sleep–wake cycle.

This is the gland that connects your hormone system to your nervous system through the pituitary gland. It is a key part of communicating when we feel pain. It also channels messages from external stimuli to our adrenals to activate the stress response. This highway of communication is called the hypothalamic–pituitary–adrenal (HPA) axis. One of the primary functions of this gland and the hormones it releases is to regulate temperature. Your on-board thermometer ensures you know when you're thirsty, hungry or fatigued and ready to sleep, directed by changes in environmental light and darkness, or by detecting slight dips in temperature. This gland also regulates the release of oxytocin – our bonding hormone.

Pituitary gland. I always imagine this gland to be a hormonal traffic conductor, as it controls so much of the flow of hormones which influence our ovarian cycle. It is the trigger gland and is known as 'the master gland' due to its major influence on the endocrine system as a way of directing key releases at specific times across your daily, monthly and life cycles.

It releases your growth and development hormones, which during puberty drive the various bodily metamorphoses we go through. Luteinising and follicle-stimulating hormones signal to your body which phase of your monthly cycle you're in. This tiny pea-sized gland is your water board monitor, supporting the flow of water and blood pressure within the body. When you have high blood pressure it's a signal of ill-health – it indicates your pituitary gland is letting you know there is pressure on your heart. It is also your on-board trigger for pain relief when you need it, as oxytocin is released from the posterior pituitary. Post-childbirth, this is the gland that triggers the highest injection of oxytocin for not just pain relief, but intense bonding with your newborn baby.

Pineal gland. This gland is the night-mode manager and is your daily sleep delivery unit. It releases the sleep hormone melatonin in response to evening darkness to ensure you have rest to recover and repair from your daily cycle. We'll explore this in more detail in the next chapter as it's a hugely important daily dose.

Thyroid. Much like a barometer, this gland maintains a fine balance of hormones that are key for our metabolism and keeping our daily energy cycles moving. It is the butterfly-shaped gland located in your neck. The two key hormones this gland releases are triiodothyronine (T3) and thyroxine (T4).

As this gland is the controller of your metabolic rate, it signals to your gut which sugars and fats to absorb, how fast to absorb them and what to convert to fatty acids. It is also involved in the regulation of your heart rate and breathing rate and is central in regulating tissue growth and development.

Metabolic activity can be disrupted when the thyroid is either overproductive or underproductive, as seen in hypothyroid or hyperthyroid disorders and Hashimoto's thyroiditis, an autoimmune disease that can lead to hypothyroidism. If you've been experiencing sudden weight gain or loss, an increased heart rate or shortness of breath, you may want to speak to your doctor about obtaining a thyroid function test.

Thymus. Much like a bouncer or door manager, this gland decides what and when external factors affect our body. This is the key gland for our immunity defences. Located in your upper chest, it is key in your armour against disease and illness. It produces hormones, thymopoietin and thymulin, which assist the development and differentiation of immunity cells. It provides extra support during puberty and neonatal phases, when your body is most vulnerable due to growth, development and changes. It is the only gland that shrinks post-puberty. This gland is the hormonal guardian of your system throughout life and offers extra help when our bodies need it. I always think it's nice to know you have glands looking out for you on the inside.

Pancreas. This gland has a dual personality, as it is also an organ. It is located behind your stomach and has important roles in digestion and energy burning. In its function as an organ, it releases digestive enzymes that help to break down fats, nucleic acids, proteins and carbohydrates from the food you eat. But in its alternative role as a gland, it magically switches from a food processor into a regulator. This multi-dimensional hero releases insulin, so it is key for regulating your blood glucose, which allows your body to use sugar for energy.

Adrenal glands. These two stress heads are the glands located on top of each kidney. This is the pair that provide us with our rapid-response and stress hormones. Along with producing adrenaline, cortisol and triggering noradrenaline, they also communicate via feedback loops with the thyroid, nervous system and the immune system to support metabolic function and immunity. We need stress hormones in the right levels to keep our body moving, but they are also highly suscep-tible to the world around us and when you're overly stressed these glands start working overtime. This is when the effects of stress begin to cause a cascade of imbalance across the other glands. Stress on the adrenals also has an impact on immune regulation and means your body is naturally more susceptible to illness. Which is why if you have had sustained periods of stress and then take a holiday, you often get ill as soon as you slow down and stop. If your body has been pushing itself to keep up with high levels of 'attack', it needs time to reduce the inflammatory response brought on by the external stress and reset immune function. If you take a break when you're running on empty, that cold or flu you've been avoiding can suddenly catch you out.

These glands are also a secondary site for the production of testosterone, the hormone which is involved in stimulating our sex drive – another reason loss of libido is linked to high stress.

Ovaries. These exclusively female glands are a pair and the two of them govern the regulation of your ovarian function, working in harmony with the pituitary traffic controller. They release the hormones estrogen, progesterone and testosterone – which as we have seen define the four phases of your cycle each month. Ovaries also regulate the key hormonal life stages from puberty to menopause.

All of these glands are responsible for key hormones and signals that keep your daily and nightly rhythms in check. They work in harmony to signpost your body's well-being so it knows when and where to send extra help. The chemicals inside us determine our daily survival. What you put into your body each day therefore has an impact on how effective these glands and their chemicals can be when it comes to doing their jobs.

OVERLOAD

When this delicate system is overstimulated with high sugar or fatty foods, stress and toxins like alcohol, we can experience a knock-on effect on how our organs and brain are working. I'm a huge believer not in cutting these things out, but being able to listen when your body needs some extra support. This is about not just looking at how your body works, but assessing the external factors and decisions you make when it comes to food and lifestyle. It's important to acknowledge what could be putting your body under extra hormonal strain and impacting on the three foundations of life: metabolism, sleep and human contact.

The glands keep the signals between organs going in a delicate rhythm needed for survival. They also adapt the rhythm, if needed, to accommodate stressors. You may well have experienced the effects of this adaptation if you drink alcohol: those feelings of tiredness, nausea, headaches etc., and the lack of meaningful sleep, are all attributable to the pressure on the kidneys or liver from a big night out, which then has an impact on all your other glands and organs. This is not to say you should never drink a drop: your body is set up to cleanse itself, and, if you have a good time, the happy DOSE hormones produced can be enhanced, which as we know has a beneficial effect. However, your glands definitely need time off to work optimally: they don't have the power to keep resetting your system without any downtime. If you don't rest your system, you begin to see the emotional and physical impacts of burning the candle at both ends.

Your body and mind are your greatest assets in achieving success – and optimising them through food and lifestyle routines is a foundation point. One of the most successful people in my life is Sharmadean Reid. Sharmadean not only received an MBE from the Queen in 2015, but has built and scaled one business, and is now on to her second, Beautystack. We have been great friends for a decade, but what has always been an inspiration to me, and what has helped her consistently achieve her business goals, is having the energy to keep going, even when there are inevitable lows. When interviewing her for this book, I asked her to share her experience and top tips for finding flow states for work: 'I first really understood hormones and their power when I became pregnant, as I learnt about cortisol. It was the first time that I realised stress could affect my baby and I needed to ensure I stayed healthy to keep him healthy. I would go to church and sing, take long walks, and began to find ways to keep calm and therefore keep my baby calm. Pregnancy was the catalyst for my realisation that what I put into my body would affect my mind and ability to love and care for those around me. This was when I realised I needed to change my lifestyle, and the impact was transformative. I stopped eating junk food, as every time I went back to that or alcohol it made me feel foggy or unfocused. Then, in 2018 when I began to build Beautystack, I discovered a metabolic reset diet that made me unstoppable. A core part of this metabolic reset was not drinking alcohol for one to three months at a time. This simple choice was transformative and the knowledge of how much energy and focus I obtained during this reset gave me power. The knowledge makes it easy to catch yourself and gives you the freedom to help yourself and unlock the energy by making simple choices about food or cutting alcohol out. It doesn't have to be forever, but it's very comforting to know you can make a choice that allows you to access so much energy when you need it most. Energy is an input and output mechanic. So, whatever you put in, will give you either more energy or take away from your energy. When running a business, it takes all your energy to multitask in the way you need for it to grow and maintain. It's a marathon not a sprint.'

What Sharmadean and I both agree on is that there is a balance in life to be struck. Being highly productive at work and having tactics for energy management doesn't mean you remove the fun from your life. Successful people multitask, but they also continually learn. To feed your brain and your body, you need fuel, energy, focus and calm. No one ever built a billion-dollar company eating badly and not taking care of themselves. You don't have to live like a nun, but you do have to understand when to employ the tactics you need for tapping into your flow states. I eat cake and pizza; sometimes I get drunk and go out late. I also work in a high-intensity environment, but I strike a balance, as I am tuned in to how and what my body needs when. I get lost in Instagram holes, scrolling through endless and meaningless content, but now I catch myself and take a break from social media for a few days, as it's a sign I need to focus on the real world and live in the present more. I travel a lot and (occasionally) make terrible relationship decisions, but I am also kind to myself about these inevitable mistakes. What is different now, post-burnout, is my knowledge of what signals and signs to look for, to help me ensure I never push my stress hormones, glands and internal system over their limits. I didn't become a wellness guru or choose to live without some of the things that make me the human I am; instead I got to grips with my patterns and boundaries.

HOW YOUR METABOLIC CYCLE WORKS

The minute food or drink hits your mouth, your brain signals to your gut that it is incoming. Chewing and taking your time to eat is key: it literally breaks the food down and the action of chewing itself produces specific enzymes that begin the digestion process. Your digestive system further breaks down the food and your metabolism is a sequence of complex chemical reactions that extract energy from the food. The energy is then distributed across the body via the bloodstream to vital organs and functions.

The energy inputs can be categorised into three groups: carbohydrates, proteins and fats. Carbohydrates are found in two forms: refined and unrefined. They come from starchy and sugary foods and are assimilated into monosaccharide sugars and used as the body's main source of energy.

Protein foods such as eggs, fish and meats are broken down into chains of amino acids and used as the building blocks of growth and repair for every cell in our bodies.

Along with carbohydrates, dietary fats are the body's main source of energy. Fats are broken down into fatty acids and glycerol which are then used to help manufacture many things, importantly steroid hormones. Fats can be either saturated – these are mostly found in animal products, and can contribute to inflammation – or unsaturated like the omega fats series (EFAs) found in plant, fish and seafoods. These reduce inflammation, which is important because, as Lola Ross puts it: 'Inflammation is our body's natural response to injury, and a protective mechanism that helps to keep us healthy. However, sometimes its normal response can become dysregulated and result in chronic or systemic inflammation. Chronic inflammation can compromise many body systems and has the potential to trigger and mediate classic inflammatory skin conditions such as eczema, as well as autoimmune conditions, mood disorders and certain cancers. When the inflammatory response is functioning normally, people often see positive changes to skin health, digestive function and are more resilient to infection.'

We need all these food groups, but not all carbs, protein and fats are born the same: there are health-giving and less healthy types, and processed and unprocessed versions. Unprocessed polyunsaturated fats from an avocado or walnuts are not the same as saturated fats found in cheese and in a beefburger. How quickly and effectively the energy can be extracted is down to the nature of the food type, not just the 'category'. It's strange that healthy foods have become luxury foods: we have engineered food to become less and less good for our metabolism process, but cheaper and cheaper. The foods that are best for our metabolism are wholefoods: the organic grains, vegetables, meats and fish found in their natural state with minimal human intervention. Yet this seems to be the food with the higher price tags.

One key process in the metabolic cycle involves the hormone insulin, which is produced in the pancreas and extracts sugar from foods such as carbohydrates. These sugars are then stored as energy to be used. However, as we all know, too much sugar is both bad for your body and bad for your energy, causing spikes and crashes. Insulin is the regulator of these spikes, but when we consume too much sugar, our insulin becomes overwhelmed and this extra energy is often stored as fats.

These reserves can be stubborn fats, meaning we find them harder to burn. As we build fat reserves we can also build insulin resilience, meaning we are not processing sugars into energy and the glucose levels in our bodies rise. This is what causes the body to then continue to store this excess fuel as fat and affects how our metabolism works. Often, a slower metabolism has a knock-on effect when it comes to our mental focus and well-being. However, as I mentioned before, our bodies require different levels of energy at different stages of life.

We have seen that our appetite levels are regulated by the hormone ghrelin, which is used to send a signal to your brain that you need more fuel, and you need to eat; and by leptin, which signals that you have had enough and quells the appetite. Eating when you're hungry and stopping when you're full sound like obvious things to do, but when your body is under extra stress or your other hormone cycles are off, you lose the power to listen to those vital start and stop signals.

When we crash-diet, we ignore these important hormones, which in turn can have an effect on how and when they're produced. The aim of your body's system is to encourage you to provide yourself with the right amount of energy needed to support its hard work each day and night. This is why when you've just completed a big workout, the endorphins and adrenaline can keep your energy levels running higher for up to two hours, and your hunger and ghrelin kick in, as your body is in need of extra fuel. This is also why we tend to crave sugar – it provides a quicker route to getting glucose into your metabolism and fast energy conversion. Having fruit and protein after working out is good; it helps your body to get ahead when it comes to providing the additional energy needed for muscle repair. As Lola Ross explains, 'Carbohydrate and protein intake are your main food goals for post-exercise recovery. Carbohydrates such as fruits and wholegrains will help to optimise and restore muscle glycogen, and eating good-quality protein such as eggs, fish, tofu or legumes will help to support protein synthesis, for the growth and repair of cells between workouts.'

Your hunger and appetite hormones, however, are not designed to choose healthy food over unhealthy food. Your brain makes the decisions to reach for the most instant sources of energy, which is often fat or sugar; and unfortunately, due to the highly addictive nature of junk food and refined sugar, we've built habits of using these foods as rewards. In return, they not only give us a sugar (and therefore an energy) hit, but release higher doses of dopamine and serotonin when the hormonal reward system in your brain kicks in. This is why when we eat cake we experience a warm fuzzy feeling.

As well as providing a sugar hit, that cake taps into all the happy memories we have of cake on birthdays or holidays. So, let's look at cake and see what happens on a hormonal level when we eat it. First, there is a dopamine and serotonin hit, as we think back to the good times before. This, paired with the ghrelin in our system, signals to us that we want fuel and this sugary delight will make us happy. Knowing that we will get a quick energy hit off it, we reach for the cake. And indeed the high levels of sugars and fats in it will cause our levels of leptin to rise, and our dopamine, serotonin and endorphins to rise too. So in the short term it gives us a high and should suppress our ghrelin and hunger pangs. However, due to the speed with which your body burns through this sugar and fat, it will not be long before ghrelin levels rise again. This means your craving for more food energy will return, and you will find yourself consuming more than your hunger hormones intended.

Remember DOSE – the happy hormones? Your body associates cake with dopamine and serotonin, in part because of the association of cake with happy circumstances. This accounts for some of the cravings: your body sees cake as a route to raised levels of serotonin, but unlike non-food-based pursuits that evoke these hormones, the cake effect is one that doesn't often give you an off switch. You crave the sweet treat not just to fill hunger, but as a way of getting access to DOSE. When you're feeling low or suffering from depression, this can lead to unhealthy relationships with food and even binge eating, a symptom of bulimia. Mental health conditions such as eating disorders and addiction are intrinsically linked to hormones, as our reward hormonal systems from DOSE. These powerful internal drugs are the antidote to low moods, but we can build unhealthy pathways to access these internal antidotes via unhealthy vehicles such as food, alcohol, drugs and sex. Cake is my biggest vice and, although I would never tell anyone not to eat it, I did find some of the hormonal backstory as to why I crave it and how my body processes it a helpful way of thinking less about my emotional impulse to eat it *all*. I suffered from bulimia in my teens and early twenties, as an effect of trauma, and it was learning about these pathways that helped me curb my own binge-eating habits. These are mental health conditions, triggered and perpetuated by the hormonal pathways we build, yet no one ever talked to me about why and how my body might be craving and leaning towards food as a medicine for pain. The lack of education on how these cycles and systems work can often contribute to the additional guilt and shame that come with these conditions.

Food is not just fuel: it is also a huge sensory memory trigger. When we eat we evoke emotions. Remember the concept of flow? I spoke to Melissa Hemsley, author of *Eat Happy*, and her description of what foods can do for us is the antithesis of chasing down a sugar-based happy high that we see in the likes of addictive eating. Not only does it show how food is nutrition for our soul as well as our body, but it is almost the definition of flow: 'Cooking and eating are two forms of self-care. When we make a meal with love and with the right environment, it can change our whole mood. Food is a ritual and can be a great way to make ourselves and others happy. It's key to find your own happy recipes, the recipes that are both nourishing in ingredients and joyous to eat. I am also a big believer in finding routines for when in the day you eat, who with, what is the lighting or the music. Eating happy and feel-good food is about listening in and working out how food can elevate your mind and body. Eating slowly, chewing properly, eating without distractions – such as eating without the TV on – these are all tips that can help your body enjoy and process the food. Eating more mindfully will add a richness to your life.'

Your body has a rough quota for how much energy it needs each day, and it needs different types of foods for different processes. Whether we're hungry after exercise or hungry as it's time for us to eat, it's important to think about what kind of food will help sustain the energy your body needs for longest. Foods that are high in fibre like fruit and vegetables will process quickly but provide sustained energy. Proteins take longer to process, which helps to steady blood glucose and keep us feeling satiated. Proteins also assimilate into the amino acids essential for growth and repair, so are super-important for post-workout recovery. In addition, these amino acids support your body to make the most of exercise benefits.

Keeping your hunger hormones healthy is also about providing your system with foods that can be best converted into the energy you need, rather than giving excess energy that is metabolised into fats and stored for a later date. Simply, too much cake as a happy-hormone reward will not feed your body with sustained energy or help balance your hunger hormones; it will trigger more hunger and mean you eat more and put extra pressure on your digestion and pace of metabolism.

When we exercise, we are creating more opportunities to burn the extra fats that may have been produced by living in a world where it's hard to avoid engineered food. Fat burning is not a process that is static: it works faster when our body needs more energy. During the menstrual cycle, this energy requirement can vary from phase to phase each month. It's also why teenagers going through puberty have much higher appetites and why women who are pregnant or breastfeeding have marked changes in their metabolism and cravings.

THE MEANING OF THE CALORIE

To help make sense of our metabolism in the modern age, humans have quantified the value of energy as a 'calorie'. This means we have stopped listening to our hunger hormones and started counting how much we're putting into our bodies. However, what calories don't account for is how quickly or effectively this food will serve in supporting your hunger hormones, to ensure you don't crave or overeat. We never talk about ghrelin or leptin: maybe these hormones just need a snappier brand name. Have you ever wondered what a calorie calculation actually is? Well I'm a geek, so I have! They are a calculation based on temperature. One calorie is based on how much energy it would take to raise the temperature of one kilo of water by one degree. Although calories are one metric or simple guide, the obsessive counting and control behaviours linked to them are not helping your body metabolise. Indeed, calorie counting can often cause more stress hormones, through the additional pressure you're putting on yourself. And we will come to see how damaging that can be.

Temperature is the key internal signal for how fast or slow your cycle should go. Earlier I mentioned the hypothalamus gland, the 'temperature trigger' that helps to regulate all the other cycles to ensure they are operating at the right pace. Your temperature represents a lot more than just how hot or cold you feel; it is literally the way your body determines its optimal operating functions.

Using your temperature is also another good way to support the boosting of metabolic rates. Studies have shown that regular ice-cold showers or cold-water submersion can enforce a raised blood temperature, which in turn can increase the metabolic rate by up to 15 per cent. This is not new science, as it's a Nordic tradition that has been practised for centuries to support health. Temperature affects our cognitive and psychological functions, which can also reduce stress. Going from a sauna to icy water is not an accessible reality for most of us living outside Finland, Sweden or Norway, but one of the things that helped me hugely when under extreme levels of stress and anxiety was taking thirty-second-long cold showers and hunting down cold-water swimming whenever possible. It's not just good for your metabolism, it is also good for reducing cortisol and strengthening immunity and improving mood. I appreciate this is not for everyone!

Another factor that can superboost your metabolic fat burning is sunlight. Sunlight has mood-enhancing effects, which in turn affect the brain's stimulation of happy hormones – serotonin, dopamine and endorphins (a natural opiate). The production of these happy hormones and our overall cycle synchronisation for day and night has been shown to have fat-burning properties. The happier we are, the healthier we are. The blue light from natural sunlight can permeate the subcutaneous fat tissue distributed across the body and breaks down these cells. Vitamin D synthesised from sunlight also optimises the absorption of calcium, the mineral that is involved in burning fat. Literally, going on a sunny holiday is good for your metabolism.

However, the sun and over exposure does of course have other negative effects on our skin, which can cause cancer. Having high factor sun lotion to protect the melanin is essential for balance.

THE EFFECTS OF STRESS

As we have seen, one of the key disrupters of our monthly cycles can be stress. In a similar way, the overproduction of stress hormones from the adrenal glands, including cortisol, will have a direct impact on the function of metabolic and hunger hormones. It is important to highlight that, although sustained periods of stress can lead to an overstimulation of compounds that can ultimately disrupt the flow of your circadian rhythm, cortisol is an important daily stimulus, there to shift our body out of a sleep state. It's our hormonal wake-up call. Cortisol, adrenaline and noradrenaline all keep us alert and focused throughout the waking day. They are essential to our body and brain's ability to move from sleep to wakefulness. But the challenge we face is that the adrenal glands that release and regulate them are sensitive to all fear and stress from external factors. This means the more external stress we are exposed to, the harder and faster these chemicals are released to keep us alert and ward off the perceived danger. So, as we build the excess chemicals released from sustained stress states, they begin to have an effect on our other core processing glands, the hormones they release and cycles they regulate.

These stress hormones can also signal to the body to store extra fats: when your body perceives itself as being under attack, it decides you may need additional energy for survival. This is why there is a direct link between women who struggle with stress or depression and weight fluctuations. Stress can also mean your kidneys, liver and gut are working overtime to detox your body and flush out elevated levels of stress chemicals. Overproduction of these can cause a backlog of processing work for your body and, as they build, they also create an additional amount of hormonal waste that needs to be removed from the body's systems; this affects filtration systems like the kidneys and lymphatic system and encourages the body to hold on to water and fats as these organs simply don't have as much capacity to cycle through the excess chemicals. The result can be inflammation, meaning women under high stress don't only see weight fluctuations, but also experience water retention and digestion issues. It can also have a direct impact on your ghrelin and leptin production, either making you feel extra hungry or leaving you with no appetite at all. Both are major signs that something is off-key.

By the time I learnt about these stress effects, it was too late and my body had hit a wall that took years to recover from. However, if we start to listen to stress and anxiety as signals of internal unrest, we can begin to think more long term about how to prevent the physical effects and protect our bodies both inside and out.

FEELING GOOD, LOOKING GOOD

The metabolic cycle is our battery power system, and optimising the glands and hormones involved can mean we not only support our insides, but also regulate the appearance of our skin and body shape. Your metabolism is keeping you going throughout the day, and when it's not working well on the inside, we see the signs reflected in a profound way on the outside.

In a society where we're obsessed with appearance, it seems crazy that we spend so much time thinking about our looks, and know so little about the outside–inside connection. One of the saddest parts of the new world of social and digital media seems to be the added pressure to look or even feel a certain way. Important conversations about body image are gradually opening up, but the anxiety of young women wanting to feel and look good sometimes makes me want to curl up in a ball and cry. Body shapes are not something we choose or pick out from an Instagram grid. They represent your genetic make-up. The beauty of your body and its shape is also defined by how you feel living inside it. When you feel happy and healthy your body glows and shines. When we feel good, we look good.

Your body shape evolves throughout your life and you can experience huge changes throughout your hormonal life stages, from puberty through to post-menopause. Your metabolism does slow down as you get older; our bodies don't need as much energy and therefore this energy system slows in line with our age and pace. Along with factors like genetics, underlying illness, diet and lifestyle, age has an impact on how fast your metabolism works: as we age our cycles and systems age with us. Cells function less efficiently than they once did, and factors like toxic lifestyles, environmental pollutants, medications and so on can put more pressure on your liver and metabolism. Yet again, stress has an impact too: as we get older we also tend to have a heavier life-load – we feel more responsibilities and more pressure from the world around us which has a knock-on effect on our homeostatic stress response, our sleep cycle function.

THE GUT AND THE LIVER

Your gut is your second brain and has the same number of neurons as a cat's actual brain. Cats are smart, intuitive and independent, so we should probably listen to our gut a little more. In Giulia Enders's book, *Gut: The Inside Story of Our Body's Most Underrated Organ*, she outlines how the gut reads every molecular and hormonal signal we have cycling through our bodies: 'The gut has not only a remarkable system of nerves to gather all this information, but also a huge surface area. That makes it one of the body's largest sensory organs.' This second supercomputer, alongside your brain, has a huge impact on the signals that tell your glands to produce the right levels of hormones needed for your overall circadian rhythm. What you put into your gut affects everything your body does.

There is an intimate relationship between the gut and liver. The gut is the first point of contact for substances that we eat and does some first checks on whether or not they are safe before filtering them into our blood. The liver receives this blood and does a second stage of checks. Often, when your hormones are off, your gut and metabolism are not working in tune to detoxify your system and remove the old hormones or bad chemicals.

Our liver function is not just our monthly hormone handler but our body's major detoxifier. As Lola Ross points out, 'Hormones have a life cycle and, once used, they are biotransformed, or detoxified, through specific phases in our liver which help to maintain healthy hormone balance. Millions of toxins, from substances like medicine, alcohol, drugs, food, heavy metals to air pollution, are constantly being handled and safely removed from our body by our liver.'

Fat stores are also part of our bodies' protection mechanism. Storage for your body when it is under strain, they are not designed to deal with modern daily digital consumption, stress and high-sugar diets. For me and many women I know there are stubborn fats, often around our middle, that can be hard to shift when under stress. There are many contributing factors to weight gain, but two things I found very interesting when resetting my own cycle and body were understanding how stress had impacted my insulin resistance and the ability to break down sugars. This in turn was linked to my liver needing a cleanse, as it had been under strain from processing so many hormonal factors and additional sugars.

Even when losing stored weight, it is vital that you keep this key organ healthy, as this is the mechanism for filtering out old stored-fat cells. Your liver is magical, as it turns fat into water and flushes it out. If your liver is oversaturated, then this process is impaired. This can happen from stress or even at more indulgent times of year, like holidays. The easiest way to shift this stubborn weight is not to go mad in the gym, but to give your liver a break. Lots can be done to help reset your metabolism – something I will come to in more detail later.

FOOD INTOLERANCE

The other big issue that can throw your hormone and metabolic cycles off when under high stress are trigger foods. Adrenal glands that are overstimulated from stress and inflammation can set off a negative chemical response in your gut, causing harm to your microbiota. Microbiota are a family of organisms, including bacteria, fungi and yeasts, which work symbiotically inside your gut, supporting the work of many body systems such as the immune, nervous and digestive processes. When toxic chemicals from stress flood your body, they can slow metabolic rates, causing mood swings and anxiety. Your microbiome doesn't just live inside your gut; there are also colonies of these organisms living inside your mouth, your vagina and your skin. The healthier these colonies are, the more effective your body is at breaking down more complex or trigger foods. However, as with any army, if you have fewer soldiers on the front line, the harder it is to assess whether this or that food is good or bad for your body. What happens, if your body cannot process and break down these foods, is an inflammation response. It can manifest itself in many symptoms, from bloating, cramps and digestive issues, right through to skin flare-ups and temperature fluctuations.

Building healthy colonies of these micro-organisms is your body's best way to process these foods. Becoming an expert and listening to your body's response to foodstuffs can help you support your colonies, by avoiding those that create stress responses. The adverse symptoms can sometimes be subtle at first, but your body will tend to get louder and louder in its signals, if something isn't agreeing with you. Don't ignore these signs; if you do, it can cause long-term problems for your overall health.

For me, one of the sadder realities when beginning to listen to my body was to pretty much remove red wine from my diet, as this is a huge trigger. Red wine has histamine in it and with every glass my face would flush, hands swell up and body would get hotter and hotter. These symptoms became more and more pronounced as I became more and more stressed. My solution at the time was to reach every night for a glass of red: my body had fewer natural defences due to my height-ened stress levels and my natural colonies became less and less able to break down the toxicity and sugars from the wine. This nightly ritual became part of the picture of how stress was disrupting my gut health and the integrity of my microbiota, which in turn limited my body's capability of processing certain foods. I began to realise it was exclusive to red wine and my body simply doesn't agree with it. So, begrudgingly and for the sake of my colonies, I now rarely drink red wine and opt where possible for natural, orange or low-intervention wine, which has less histamine and doesn't evoke this extreme physical response.

If you suffer from conditions such as irritable bowel syndrome (IBS), or the overactive immune system seen in conditions like psoriasis, you can develop intolerances to foods, even ones which appear to be benign 'natural' substances. Prolonged stress, alcohol consumption, and many medications including the oral contraceptive, can diminish the health of your gut ecology. This can lead to dysbiosis which compromises the integrity of your gut lining and results in less selective filtration of your broken-down food molecules. Such 'perforation' (known as leaky gut syndrome) can allow larger food particles into the bloodstream, prompting it to send an alarm to the immune system which sees the food particles as an invader rather than food. And this is how, through imbalances or overstimulation of certain systems, we can develop intolerances to foods. We often discover these by simply eating particular foods, which results in them not 'agreeing' with us. This can cause any number of phys-ical symptoms, commonly bloating and other digestive issues, but intolerance can also present as skin issues, concentration and mood disorders. However, what we often aren't made aware of are the emotional signs, such as anxiety or fatigue.

When we put something 'toxic' into our bodies, we receive a hormonal trigger from the adrenal glands, telling our brain this food is bad. When these signs are ignored and we continue to eat triggering food, we experience higher levels of internal stress hormones and put our liver and kidneys under more pressure, as they work harder to cleanse our body. These symptoms can become a disruption to our metabolism, sleep cycles, or both.

Our hormones cause inflammation as a reaction to imbalance. Have you ever cut your hand and been fascinated as your skin magically heals? In the first stage of this healing, throbbing pain and blood are part of the inflammatory process followed by the formation of a scab for the repair. However, if you continue to pick off the scab the cut gets deeper and more gnarly. In just the same way, ongoing negative stress or continued consumption of trigger foods can cause a cycle of inflammation in the body; and often, rather than reduce stress or remove the trigger, we carry on as normal, never allowing ourselves to get past the inflammation stage and have the opportunity to heal.

This is a response triggered by a dominance of certain hormones, such as cortisol, which in turn creates more internal defence responses such as inflammation. Internal inflammation can often be signalled by external swelling. It is at its most obvious when we have an allergic reaction to something, but it can be a subtler swelling around the glands, including the neck and face. Hot flushes or skin irritations are also a clear sign that something is not right inside. Water retention is another sign, which was one of my big issues with stress. I would swell up like a balloon when I had soya, red wine and caffeine. Rather than listen to these warnings, I pushed through, but over time my internal 'wound' became extreme, my metabolism slowed and I gained weight.

There are some simple tricks for identifying if you have trigger foods in your diet: the elimination diet is still the gold standard of testing for intolerance to certain foods and it is free. The steps are to identify the food you want to eliminate, and remove it entirely from your diet for twenty-one days. You then reintroduce that one suspect food on day twenty-two – one serving in the morning and one in the afternoon – and monitor any mood or physical symptoms carefully, making sure you record any changes. If you have removed several culprit foods, reintroduce each one separately so that you know which might be causing issues or not.

Trigger foods tend to sit within specific food groups linked to specific hormones. Each impacts your body's hormone levels and alignment differently. *The Hormone Reset Diet* by Sara Gottfried is a brilliant book that transformed my understanding of the connection between foods and hormone groups that can trigger hormone imbalances. There are seven pillars that she highlights and they each connect to specific hormone responses. Here are six of them that I have found particularly useful:

Meat and soya → estrogen dominance. With so much of modern farming involving synthetic hormones such as estrogen, it's no surprise it's making its way into the meat you eat. Soya foods, meanwhile, contain phytoestrogens – substances that have a similar molecular structure to our own endogenous estrogen and which can mimic and affect estrogen behaviour in our body. High consumption of certain types of soya products may be linked to estrogen dominance in some women, a hormonal state which can give rise to a range of effects on cycle health, mood and metabolism.

I went through a soya latte phase and it was only when I was experiencing nausea and early-pregnancy symptoms that I realised something was wrong. When I finally spoke with a nutritionist, I made the possible link between my symptoms and overzealous consumption of soya milk.

Sugars and glucose including fruit → reduced leptin and higher ghrelin levels (AKA you're always hungry). As we have seen, leptin and ghrelin each have opposite roles in regulating appetite. When we consume too much sugar, it causes our appetite regulation signals to be thrown off by the stimulation of sugar-highs. Removing sugars for a period of time can effectively reset these hormonal signals. These hormones also affect how easily our body can convert glucose and sugars into energy or fats. Erratic appetite is a sign these hormones are off-balance; too much sugar can perpetuate this imbalance and mean your body stores the surplus food as fat, preventing your natural hunger hormones from signalling when you're full.

Caffeine, alcohol and refined sugar → cortisol. When we're stressed we run on adrenaline. Often, when we crash after the highs or lows, we reach for things that give us energy. This is commonly caffeine and sugar, which can become highly addictive, and it is easy to get stuck on cyclical energy-fix behaviour. Like a roller-coaster ride, you get used to the 'sugar-high to low-energy' slump and reach for more fast-energy food, which causes the cycle to begin again. Beware, as this is 'fake' energy. Combined with the stress hormones you're experiencing, it is what stops your body from being able to make new energy, meaning it quickly burns through each temporary boost and turns the surplus materials into fats. This drives inflammation and stress levels up. Our cycle of stress can be perpetuated by the cravings that come when we feel we have to 'push through'. What we should be doing is stopping and resetting. If you 'need' coffee, it's a good time to quit and reset your system. No one needs coffee to get started in the morning or as a midday pick-me-up. It's nice to have and a lovely indulgence, but it shouldn't be a crutch.

Grains → thyroid. This doesn't just mean bread, but also oats and all forms of gluten. These products can often raise our blood sugar levels rapidly, especially in their refined versions (the brown husks in wholemeal grains contain fibre that slows down the sugar released from the inner grain). Additionally, grains often contain gluten – the protein found in the inner part of the grain anatomy. This protein has a glue-like texture, making it troublesome for some people to digest, and it can aggravate many conditions. It takes longer for your gut to process due to the specific enzymes needed to break down the gluten. Your thyroid can be affected by the long-term impact of this overstimulation and therefore under- or overproduce TH levels. Giving your system a rest for five to thirty days can reap benefits for your thyroid and digestion.

Dairy→ **growth and digestion hormones**. This is not just about weight. The saturated-fat content in dairy products has inflammatory actions in the body. Aside from saturated fats being harmful to cell membranes and reducing cell health and communication, dairy can cause an overproduction of mucus, leading to blocked sinuses and, in turn, limited breathing and oxygen absorption. Oxygen is key not just for digestion and metabolism, but also for your brain, focus and general sur-vival. We have built a culture of flat whites, but never been told that the milk in our daily coffee could have such an impact on our stress levels and ability to breathe. A recent cultural focus on veganism and mass dairy farming has encouraged more awareness and opened up more options such as plant milks.

I love my daily ritual of coffee on my way to work. However, now, when my stress is high, I've switched my morning coffee for a green tea and have coffee as a treat, when my stress levels are in check. The green tea still gives me a morning caffeine boost; the L-theanine within it helps sharpen my focus and it also has properties that actually help support my metabolism and digestion, meaning it has a mood-boosting effect. The good news here is that, like skin, your hormones and cycles have powers to repair, so if you realise you have an intolerance and you suspect which group or specific food is your trigger, you can take a break from eating it for a while and reintroduce it when your body and cycle have had a chance to readjust. This doesn't work every time, but especially in cases where these triggers are linked to stress, if you reduce the stress-induced inflammation, your body may be able to process this food again. Your hormones have a magical way of realigning, but they sometimes need to be given the headspace and time to cleanse out old patterns to make room for a new harmonious system.

A 21-day food reset can support liver function and reboot your blood sugar balance. Cut out alcohol during this time and support your phase-one detoxification processes with plenty of dark green vegetables and superfoods like broccoli sprouts which contain indole-3-carbinol compounds. Sulphur-containing foods are also great for liver support, so include daily servings of brazil nuts, eggs, kale, onions, garlic and cabbage. Supporting the gut microbiome with prebiotic and probiotic foods is also crucial, remembering that our ecology of gut bacteria plays a role in liver-detoxification processes.

The other key to keeping a happy and healthy liver is to eat high-water-content vegetables and fruits and drink around two litres of water each day. This supports detoxification and helps fat burning.

Your metabolism is not checking how nutritious, healthy or triggering the foods you eat are. It doesn't care if you're going vegan for January or doing Sober October: it cares that what you're eating is supporting your body's ability to function and metabolise food to fuel for every day, week and year of your life. In return, it will give you happy hormones, mental health, focus, energy and a balanced sex drive. When you're listening out for the moods, symptoms and cycles that signal your hormones are off, you can course correct and give them a chance to reset, to get back to your happy flow.

SUPPORT FOR EXCESS HUNGER

Hunger, much like all our moods and symptoms, is a signal for a function we need to survive. Eating is an essential part of our systems. However, our bodies were not designed to live in a world of such abundance. This means we can often throw off the natural balances of our hormones indicating a need for food, as we have it all so readily available. It is healthy to feel hungry, but what often happens when our hunger hormones are out of whack is that we begin to feel hungry all day. Having an 'insatiable appetite' or craving junk food is also a sign your stress hormones are high. Junk food is often a reward and cycles of hunger or craving foods that stimulate an immediate sugar hit can be hard habits to break, especially when we're under a lot of stress. Ensuring that your diet contains plenty of good-quality proteins each day is crucial for production of all hormones, including ghrelin. Healthy polyunsaturated fats, especially EPA and DHA from fish, sea vegetables and omega-rich plants, can improve hormone communication across the entire endocrine system. As your microbiome is so important to your body's overall function and processing power, you can support your metabolism by supplementing a daily prebiotic and probiotic after food. Supplements that can also be taken if your hunger isn't subsiding after eating include L-glutamine, chromium, cinnamon and green tea.

Protein, fats and fibre can help to stabilise blood glucose which may curb cravings and hunger. A scoop of protein powder in water or juice or a teaspoon of nut butter can be used as a post-meal cravings fix. Flaxseed powder is a soluble fibre which has a swelling action in the gut and can help to manage glucose levels, reducing hunger. For stabilising longer-term cravings, you can try the trace mineral chromium – known for its positive influence on insulin and glucose regulation.

NO APPETITE

When we're stressed or going through pain, leptin and adrenaline hormones can force our body into fight-or-flight mode, meaning we miss almost every signal, as we're mostly just surviving each day. Eating is essential for your body's energy; it is also helpful to support sad and stress hormones. What to eat, when to eat and why is important to understand not as a trend, but to help make sense of the signals your body is giving you and how to support it. Treating food and a meal as a moment in time for self-care, making a meal and setting the table is a great way to kick-start a more emotionally positive relationship with the food you're eating. This applies especially when appetite is suppressed from stress.

HELPING YOUR HORMONES RESET

As mentioned above, sometimes if you're still feeling the effects of hormonal imbalances, it can be helpful to employ a reset through regulating your blood sugar levels, by using methods such as intermittent fasting. This method should allow your body time off from processing food and after a few weeks give you a more intuitive eating pattern – one where you eat when you're hungry and not when you're not. This intuitive eating is by far the most natural and optimal way of operating, which, when you think about it, is fairly logical. Prescribed mealtimes is more a social structure than a system designed for our body's balances.

AVOIDING HORMONES HIDDEN IN FOOD

Processed meats and dairy are bad for your metabolism, not just because they serve less nutritional value, but because cheap or low-quality meat also contains both synthetic and natural hormones. The animals are often medicated with artificial hormones or, in the case of dairy cows, they contain the mother's endogenous hormones naturally. This is then passed on to you and can contribute to endocrine disruption.

The liver can be thought of as a chemical-processing zone. It is crucial in handling and removing toxins and waste safely from the blood, which are eventually excreted via our urine, sweat and faeces. The liver relies on antioxidants such as glutathione, which we obtain from our diet (plant foods), and manufactures more cells with them to metabolise toxins. You can find supplements such as NAC which support glutathione production.

Supporting the health of the microbiome is now known to be important in relation to its role in waste-hormone removal. If used hormones are not eliminated from the body they can re-enter the circulation and lead to hormonal imbalance. A good-quality Lactobacillus and Bifidobacterium probiotic supplement can replenish your levels, and 'feeding' your existing microflora with prebiotic fibre powders may be even more beneficial in supporting the gut bacteria. There is a vital process within detoxification called methylation which is crucial in heavy-metal and hormone handling; this relies on large amounts of folate, B6 and B12 which you would get from wholegrain, leafy vegetables and meats, but supplementing with a methylated B complex can be a really useful support in detoxification for many women.

What is clear for everyone is that food is fuel, but not all food is processed by our bodies in the same way. Thinking about your own body's balances will determine what works for you. As with all these cycles, your metabolism is a system that is unique to your own pattern and when you tune in to it you can feed it in the right way and tap into the magical flow that comes from happy and healthy cycles.

Finding your flow is not just about learning how your body works, but also listening to what your body is telling you.

TAKEAWAYS

- Metabolism is part of your 24-hour circadian rhythm, and the hormones involved in your day contribute to how well or badly you metabolise food.

- Hunger and appetite are hormonal signals. The hormones involved are: ghrelin, which signals you are hungry; leptin, which is released to signal you are full; and insulin, which regulates your metabolism.

- There are eight key hormone glands across your body and each releases hormones throughout the day and night, which in turn steer your body from day to night.

- Food is fuel for your hormones and metabolism, but different types of food and supplements can help elevate or prohibit your body's energy and metabolism.

- The happy hormones of DOSE will help your body metabolise excess fats and sugars.

NIGHTLY RHYTHMS:
SLEEP

It is no exaggeration to say that sleep is essential for staying alive. Sustained sleeplessness leads to physical and mental disruption. Sleep isn't a nice-to-have, it's your body's baseline for survival. The reason it is so necessary is that it allows for the nightly hormonal reset that means your body can recover from the day's action.

Never before in civilisation have humans had more stimulation in the day – from rich food, pollution, screen time, stress and the fact that all these are available 24/7. In the US, it is estimated that almost 45 per cent of adults are affected by inadequate sleep. Our days are packed, and if you don't get sleep, your body simply doesn't have time to detoxify, ready to start all over again. Your metabolism is your day cycle for energy production, to push you through whatever you throw at your body and mind, but without sleep, the whole orchestra falls apart, as nothing has time to reset, ready for a new performance. In his book *Why We Sleep*, Matthew Walker describes the resetting we gain from sleep as an 'autobiographical sculpture of stored experience'. As he puts it, due to the ever-changing nature of life 'our brain always requires a new bout of sleep and its varied stages each night so as to auto-update our memory networks based on the events of the prior day.' Walker's book has changed so many people's lives, by sharing the science behind this essential life cycle. Sleep is not just important to heal and reset our hormonal systems, but also helps to heal our minds and memories. Our brain processes and resets during sleep cycles and, to help heal, we need to dream. Neuroscientist Dr Tara Swart outlined a phenomenon that occurred during the globally enforced lockdowns caused by the coronavirus pandemic of 2020 that demonstrates just how dreaming interacts with chronic stress and anxiety: 'What we have seen more recently, since lockdowns and macro stress, is this trend in people experiencing the vivid dreaming phenomenon. Vivid dreaming is actually psychological processing of complex emotions. The present global uncertainty, and background chronic stress, is causing people to have anxiety dreams, but the content of the dream is connected to historic anxiety memories. However, although these dreams may be unsettling in the moment, it's actually a good thing that people are dreaming. Dreaming allows us to process complex emotions and is why it is so important to get good sleep, as this will help heal your anxiety and stress through processing it within dreams.'

As with metabolism, sleep is triggered by a hormonal chain reaction and relies on temperature – two things that fluctuate throughout women's menstrual cycles – so how deeply you sleep can be affected by where you are within your monthly cycle.

So, what are your hormones doing as you sleep? The first thing to understand is that there is a sequence to your sleep and each phase has a different effect for your nightly recharge. These sequences are ninety minutes long and vary between NREM (non-rapid eye movement) and REM (rapid eye movement). The muscles in your eyes flutter faster or slower, signalling to your brain to cycle from one sleep phase to the next, as we alternate between light and deep sleep.

The interplay and alternation of NREM and REM phases is dictated by time of night, temperature and functions happening inside our body. REM is when we dream; it is also when our brain releases oxytocin, one of our happy hormones. There are roughly five steps our body goes through from a wake state into our first REM, and each has a hormonal trigger. From hitting our first REM stage, we then alternate between REM and NREM stages across five phases throughout the night. This is what Walker describes as sleep architecture.

The hormonal steps that kick in before this first REM phase are all part of the nightly recharge:

Stage one. Eye movement and muscle activity slow down. Melatonin begins to rise. Melatonin is released from the pineal gland in the brain and that stimulates the release of gamma-aminobutyric acid (GABA), which is a neurotransmitter. The combination of GABA and melatonin is your sedating duo (just as the relationship between serotonin and oxytocin is your elevating duo). They put your brain and body into peaceful and repair state during sleep. They work in unison to relax the body, helping you both fall into sleep and stay in a sleep state. Low levels of both melatonin and GABA cause insomnia and these two calming chemicals are the body's essential sleep aid.

Stage two. This is still a light-sleep state. Eye movement stops, brainwaves and heartbeat slow. Our body temperature drops and this is a signal to our brain and body to begin the hormonal reset. Due to women's temperature being affected by rising estrogen in the first half of the monthly cycle, it can be harder to slip into sleep in this phase, but in the second half of your cycle when you have higher progesterone, this can mean you have deeper sleep. Tracking your sleep cycle, alongside your menstrual cycle, will give you indicators if these rising and falling chemicals are affecting your nightly rhythms.

Stage three. This is a deeper sleep state, the time when replenishment takes place. Growth hormones and prolactin hormones are released from the pituitary gland for internal organs, bone and tissue repairs. These hormones contribute to immunity and metabolism cycles too.

Stage four. Blood flow and prolactin hormones mean your physical energy and muscle repair happens in this stage. Like a recharging battery, you need this phase to ensure that your levels are ready for a new day of energy consumption.

Stage five. This is the REM stage. Your heart rate and temperature will increase slightly. Muscles become temporarily paralysed. Oxytocin is released and your dream state is triggered.

We cycle through these five stages several times per night. There is a common myth that everyone needs eight hours' sleep. Most research now shows that people's needs vary from five to nine hours, dependent on our genetic make-up (there are genes that regulate our own unique sleep–wake cycle), which accounts for early risers and those who naturally rise much later. The best way to tell how much sleep you need is simply assessing how rested you feel in the morning.

What is key to sleep is that it provides essential time for your body to reduce cortisol and adrenaline levels, brought on by the day's stress. It is during this time that your body builds essential defences and immunity. During sleep your body releases not just a series of hormones, but proteins such as cytokines, which act as the first defence against disease and infections and fight internal inflammation. The healthier your sleep, the healthier your immune system and body.

The magic of sleep is known by us all: no one can deny the incredible sensation you feel after getting a good night's rest. What I find most fascinating is that, although this high is actually a sign that our body is working effectively, if we wake with a groggy or sad feeling after a bad night's sleep we tend to try and ignore it. We often resort to pushing through and hoping the next night will bring more luck. However, pushing through in these circumstances almost inevitably leads to the production of more stress hormones to try and get you into the next sleep cycle. And an increase in stress hormones has the capacity to reduce the stimulation of melatonin, making the problem worse.

Sleep also allows your sympathetic nervous system a rest. One responsibility of the sympathetic nervous system is that it transfers fight–or–flight signals from the brain to the hormonal body.

It was only in the twentieth century that scientists were able to connect the dots for why humans need sleep for cell regeneration. If you think about it, sleep seems like a very vulnerable state and not very logical for a mammal like humans: when we sleep, we have no alert or defence mechanism. What is now known is that sleep is essential for the nightly repair of our body's cells and specifically our autonomic nervous system (ANS). This system is like our internal fibre-optic internet connection. It is how our body moves and feels.

There are three different parts to the nervous system that makes up our internal internet: sympathetic, parasympathetic and enteric. Each controls responses across a different part of our body. Sleep is essential for the rest and recovery of all of them, so that they can keep our bodies alive, alert and responsive each day.

The parasympathetic and sympathetic nervous systems control the activity of cardiac cells, smooth muscle cells, immunity and endocrine glands for hormone release. The enteric nervous system (ENS) exclusively controls the function of the digestive system.

Our sight is a key sense for telling the ENS how to respond. Have you ever cut your hand and found it was only when you saw the wound that the pain kicked in? This is your sight forcing a nerve pain response. Your eyes are the central part of sleep, not just during NREM and REM cycles, but also for identifying daylight and night-darkness cycles. Darkness is a signal to your pineal gland to release melatonin and trigger GABA, but it also signals to your ANS. It tells your nervous system it's time for rest and relaxation. Which in turn also relaxes muscles across your body. It is time for your body to begin its nightly recovery.

Changes in ENS activity such as relaxation of muscles occur on passing from wakefulness to sleep and then also within sleep states from NREM and REM. Research shows there is more repair during NREM states, which is linked to increased oxygen intake due to heavier breathing in NREM, with lighter, more shallow breaths in REM.

As we all know, oxygen is essential for us to stay alive, but breathing is often overlooked as just something our bodies do. However, proper breathing is one of the most powerful assets you have to support your body's health. Often, when we are under greater stress in the day with higher cortisol and adrenaline levels, our breathing can become shorter and shallower, taking in less air on each breath. This for me could sometimes lead to a panic attack. These attacks are mental and physical and are the body's way of showing us we have hit a limit. During an attack, it can feel as if you're drowning. It is as though there is no way of getting air and the panic feeds more panic. The remedy for my own panic attacks was learning to regulate my breathing, adopting methods to clear my mind and take long, slow breaths. The response is 100 per cent effective for me and means that, within ten minutes, I can take myself from drowning to swimming through life again.

Learning to breathe properly is also a huge asset for those who struggle with slipping into sleep. Doing deep-breathing exercises before bed will not only help regulate and support the release of sleep hormones, but also relax your ANS, so that your body signals it is time for sleep. What this means for our hormones and nervous system is that they regulate in response to more controlled oxygen intake. When we sleep our body naturally creates varied breathing patterns between NREM and REM states.

CHEMICALS INSIDE US KEEPING US AWAKE

Have you ever lain awake at night while the small or big things swirl around in your head, keeping your mind and body from slipping into restful bliss? As you are probably only too aware, the cruel mental panic that begins to kick in when you realise you can't get to sleep in itself causes stress hormones such as cortisol and adrenaline to release. This in turn keeps your brain alert, prolonging your wakefulness. You can see the hours slipping by and still you lie awake with your destructive thoughts and stress hormones swirling, and so the cycle continues. This for some people is a personal prison and nightly torture.

You will be pleased to know that, like all stress-induced problems, the hormonal responses that form part of this vicious circle have internal antidotes too. When you understand the internal chemicals, you can begin to find ways to reduce the stress hormones and sleepless cycles.

As we have seen, melatonin is the master hormone for the transition from wakefulness to sleep. This hormone is released from the pineal gland in your brain. The pineal gland is directly affected by light levels. Dark or low light switches on the gland and can trigger the release of melatonin. This is why, if you are overexposed to bright lights from screens throughout the day, it has a direct impact on how effective your pineal gland is at kicking in and releasing melatonin. We live now in an era of synthetic light and it is having not just an effect on our sleep, but on all the other hormones that need sleep to reset.

OUR MONTHLY CYCLE

As we have seen, the stages of sleep can be affected by the ovarian hormone cycle. This is due not just to the changes in our hormone production during each phase of the month, but also to the temperature fluctuations that happen throughout the phases. Life stage plays a part as well: when young women are going through puberty, they tend to sleep almost three to four hours longer than a woman in midlife. This is in part due to the amount of hormonal growth and development that is happening inside her body at this life stage, but also because once women have a menstrual cycle, their estrogen cycle regulates around rising and falling temperatures each month. So for women in midlife, sleep can be affected depending on where

STRESS RESPONSE SYSTEM

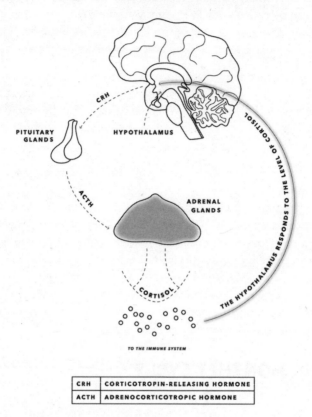

CRH	CORTICOTROPIN-RELEASING HORMONE
ACTH	ADRENOCORTICOTROPIC HORMONE

in your cycle you are. It is also why sleep is disrupted for women in perimenopause, as this is a time when body temperature can fluctuate even more aggressively and sporadically. Women post-menopause often report far more regulated sleep patterns, as their hormone levels have cycled through the monthly phases and become set into their second stage of life now they no longer have an ovarian cycle.

For women with monthly cycles, the increasing estrogen levels during the rise phase can create hyper-focus in your day, but for some it can also mean the additional alertness feeds into a busier brain, thus making it harder to calm thoughts and slip into a restful sleep. In other words, during this time of the month it is more important than ever to have rigorous sleep-hygiene routines. Conversely, during the fourth or luteal phase when progesterone – a sedating hormone – is higher, some women find they have much deeper sleep. This is why tracking your sleep across your cycle helps build a picture of what your hormone levels are doing.

Estrogen dominance can be another blocker to good sleep in the PMS phase, not just from the disruption and pain it can cause, but also the impact it can have on the other vital organs for supporting overall hormonal balance. Excess estrogen can be linked to disrupted metabolism and poor sleep: if these two cycles are not working effectively, your body struggles to excrete estrogen from the body. It is the liver that supports the processing of estrogen both in the day and at night. Stress on the liver from poor metabolism cycles and sleep cycles therefore is a major contributor to high estrogen levels.

PMS symptoms are very commonly linked to estrogen dom-inance. This is another reason why I avoid alcohol in the fourth phase of my cycle, before the onset of my bleed: to reduce overstimulation and associated later nights. As we have seen, when I drink during this phase my body does not process the toxicity as effectively via the liver. I find it is also the time when alcohol evokes greater anxiety in me, due to my higher level of progesterone and the emotional awareness and sensitivity brought on by this hormone.

So it is perhaps unsurprising that, when I don't drink in this fourth phase, my PMS symptoms almost disappear and my bleed becomes far more regular, along with my focus, energy and sleep. This is something we have been researching across the Moody community and have found women report back that removing sugar and alcohol during this phase had a transformative effect on cramps, bloating, fatigue and even heavy-flow reduction.

The positive by-product of this monthly temporary pause on my favourite wine at the end of a busy day is that my nightly rhythms regulate and my moods and symptoms become more balanced. This does not mean I stop drinking wine or indulging in doughnuts occasionally, but I choose when in my month my body is going to be best at managing the extra sugars, fats and fatigue. These choices become a habit and in turn build a foundation for good sleep hygiene and daily health routines.

What happens each night has a direct impact on what happens each day and therefore, over time, how effectively you are able to operate in life, work and relationships. Better sleep means my overall health and well-being have improved as well. I now don't get ill every winter, which is a sign my immunity is boosted, and on a more immediate level my ability to navigate hard relationship challenges and the general day-to-day have become effortless. The easiest and simplest way to help regulate these cycles is by tracking to help you build healthy rituals.

SLEEP AND APPETITE

Sleep is also important to reset your hunger and metabolism hormones, for optimisation of energy production in the day. Sleep helps control insulin, as this is not being spiked by us consuming foods or drinks during this period. This in turn reduces adrenaline and cortisol and allows the body to return to its homeostatic levels. Insulin and glucose balance are essential processes for ensuring that when your body is awake and alert, you have the energy you need. When you are awake your body draws on insulin to provide energy to function by turning blood sugar (glucose) into energy. You need more energy while awake as you're physically moving and using up energy both stored and consumed each day from food. Energy is still being drawn from your body while you sleep, but crucially this energy is not disrupted by the spikes from food being ingested. So the energy can be focused on specific repair work such as restoring your nervous system. While you sleep, on average your blood sugar will surge at certain points – often around 4 a.m. to 8 a.m. This surge is then managed by muscle, fat and liver cells absorbing and processing additional glucose.

In simple terms, your insulin levels are like a spirit level: your body is continually trying to keep a balance and your sleep cycle is one key part of how your body can regulate these levels without you consuming more sugars from foods, which will have a knock-on effect on both the production and effectiveness of the insulin in your system during the day.

When your body is under too much stress in the day and is therefore producing more cortisol and other stress hormones, this also has a knock-on effect on your insulin production during both day and night. When you hit high stress, your body can begin to block insulin, causing your blood sugar levels to rise. This in turn means you store more sugars and fats, and your sleep reset becomes even more essential. However, the cruel reality is that the raised levels of cortisol and adrenaline that stress provokes tend to keep your brain alert and prohibit your ability to get the much-needed recharge.

Aside from regulating our appetite, the hormone leptin has a role in energy burning. Melatonin communicates with leptin at night so if we miss vital sleep hours this communication breaks down and it can be much harder to shift weight.

Understanding the role of sleep in weight management helped to piece together the puzzle in my own journey. The stress I was experiencing in the day meant I was either having disrupted sleep, no sleep or to get to sleep I was drinking red wine and other sugar stimulants at night before bed. Of course this meant my body was not able to remove the additional sugars from the wine, meaning my sleep was disrupted by the alcohol and sugar, and the excess sugar was stored as fats. All this, combined with my interrupted melatonin–leptin signalling, meant that I found shifting weight almost impossible.

It also meant I would wake up from sleep feeling more tired and fatigued than the night before. The stimulating effects of adrenaline and cortisol had worn off overnight and I was left to pick myself up each morning and try and find energy from what felt like an empty tank. This cycle of building stress in the day from adrenaline and cortisol, consuming stimulants before sleep and even when having sleep feeling exhausted was the pressure cooker that meant my body stopped producing the hormones connected to my menstrual cycle.

What I was doing was pushing my body so far that increased stress-hormone production was depleting my thyroid function, which in turn led to a hormone communication breakdown to my hypothalamus and ovaries: my periods totally stopped. Remember the hormonal triangle we saw in the last chapter? That delicate interconnectedness of the endocrine system is a dance of hormones between the three key axis points, adrenal–thyroid–ovarian, when you hit higher stress cycles and hormones from your adrenals, you then have a knock-on effect to thyroid and ovarian cycles, as your body tries to recalibrate and balance.

This, frustratingly, is a very under-researched area of women's health, even though seemingly a fundamental part of the human experience for 51 per cent of the population. I know this triangle was at play, due to the fact that my periods returned when I was able to regulate the stress levels affecting my thyroid and adrenals. My sleep cycles themselves became regulated, which in turn contributed to the reduction in stress – a virtuous circle. However, I still have some male doctors tell me that stress levels and a loss of periods are not connected, as there isn't enough research to support this 'theory'. My simple answer to anyone who holds this argument against women and their experience of hormone cycles is to point out that the only reason this hasn't been researched effectively is gender inequality. Thankfully, a majority of modern scientists can see that, based on the simple way in which our hormone balances work, there is of course a direct correlation between stress and disrupted hormone cycles such as menstruation and sleep. My point here is not to dismiss the amazing research and work being done in women's health now, but to highlight why this important relationship between our internal chemical balances is not more strongly emphasised.

The other variable that can hugely impact our daily and nightly rhythms is travel across time zones. If you have ever experienced jet lag, you will know that one of the most frustrating aspects of this destabilising state is the body's misalignment with sleep. For me, the worst part of my burnout, partly fuelled by travel across the world for work, was becoming trapped in a permanent state of jet lag. The symptoms include dizziness in the day, hunger and appetite fluctuations, dehydration, mood swings and struggling with getting to sleep followed by waking up tired. These are all driven by the hormone disruption created by playing with our sleep–wake cycle. They are also the symptoms experienced by those who suffer from insomnia or ongoing sleep disruption: it quite literally makes you feel like you are permanently lagging behind the rest of the world.

If you don't help your hormones get back to the right day and night cycle, your internal system will slip into fight-or-flight and again use adrenaline and cortisol to push through, so if you are often in and out of different time zones, it can have a huge impact on your endocrine system's natural rhythm. Which in turn has immediate impacts on metabolism, immunity and general mental and physical health.

EXTERNAL CYCLES FROM DAY TO NIGHT

Our bodies are essentially clocks, operating not just around cycles, but around daily and nightly rhythms. The energy that keeps these rhythms in time comes not just from the food we consume or how much stress we are exposed to, but also from our environment and other factors such as how much light we receive in the day, or how little, subject to the season and where in the world we live.

Daylight and sunshine are nice, but more than that, they play a huge part in how effectively our bodies process hormones at night. Your happy hormones, including serotonin, are stimulated in the day by sunlight and warmth, which in very simple terms means that people living in countries and environments with longer days have more exposure to healthier rhythms and cycles. However, higher temperatures at night can mean it is harder to slip into a natural sleep cycle, as it is a dip in our bodies' temperature that triggers melatonin for sleep. Conversely, studies have shown that in countries such as Norway, with extended winter or darker seasons, there is an impact on people's mental health, meaning that although temperatures are lower at night so sleep is easier, the lack of daylight and lower serotonin production can impact on how deeply people sleep.

Our body clocks are also dictated by the very location and environment in which we live. Women living in cities are more exposed to higher pollution levels, stress levels and general hormone disrupters than women living in more rural areas. Our society's shift towards urban living is having an impact on our hormones; so when you are tracking your sleep, tune in to the external stimuli you are exposed to. This can give you clues as to what triggers are potentially affecting your hormones and causing disrupted sleep patterns.

Your circadian rhythm is the foundation for understanding your body's day and night literacy. This rhythm dictates your sleep patterns and Matthew Walker states that almost 30 per cent of people are actually operating on a 'night owl' clock, meaning they are not morning people. What is also true of our body clock is that it is affected by weather. You are the best judge about whether you are a morning person or a night owl, based on how effective your focus, energy levels and general moods are each day. The more we can understand how all these factors affect our internal chemical balances, the more informed our choices become. The aim for us all is to become more proficient in the language of our body and the rules by which it operates, so we can then help build a stronger and longer-lasting system.

The point of paying attention to your hormones, as far as sleep is concerned, is that by doing so you can develop a sleep routine that suits you. I must caveat here that these routines and rituals are not aimed at people suffering from more extreme symptoms, such as chronic insomnia, which can sometimes be related to more serious underlying conditions. If you find yourself in this category, then you should seek professional medical advice – though if you do so, you will find that tracking your sleep and your hormonal rhythms will give you valuable information.

RHYTHMS, ROUTINES AND RITUALS

There is sadly no magic pill to solve long-term sleep issues. Pharmaceutical drugs can deal with the immediate issues of getting to sleep, but these sedatives have often been linked to addiction and in some cases dependency. Living on sleep medication for life is for many people a daunting and some-times debilitating reality. If possible, it is far better to try and address the underlying issues, even if this is unlikely to be a quick fix. Sleepless patterns are often linked to emotional root causes that occur during your wake cycles. Maintaining logs and journals about what types of thought patterns creep into your mind and keep you awake is one way to start identifying where this disruption may be coming from. Sleep, as with metabolism, is about the rhythm your body needs for homeostasis. Finding that rhythm isn't easy for some people, but it is the first step to working out how you can support your natural cycle and encourage the hormones involved in this essential function.

There are also huge benefits to building a routine of breathing exercises, as this rhythmic breathing helps keep the oxygen-ation of your body clear. Simple truth is oxygen is key for survival and we are often forgetting to breathe properly, especially when we are stressed. There is a brilliant exercise called belly breathing, a technique I learnt from our Moody breath-work instructor Natalia Sketchley whose Instagram @yoga.sketch is dedicated to breathing techniques. There are more intense practices that you can build into, which can sometimes take people into very deep physical and mental regeneration states. Belly breathing is my favourite daily practice and involves three simple steps. Try it, and you will feel your whole body and mind calm, as the process automatically connects your brain and nervous system.

1. Sit or lie flat in a comfortable position.

2. Put one hand on your belly just below your ribs and the other hand on your chest.

3. Take a deep breath in through your nose, and out through pursed lips (sounds like a hiss). Focus on pushing your belly out as you exhale. Feel the movement of your belly with your hands and repeat that rhythmically ten times.

Repeat this exercise on a daily basis, even if you feel good, and it will help support your nervous system and sleep cycles long term.

To help build a routine that works for you, you need to under-stand how much sleep you require and your personal sleep patterns. Track your sleep, in particular reflecting on nights where you felt rested when you woke up. Look at when in your monthly cycle you sleep deepest and dream compared to when you feel restless and wake with a busy brain. When you have tracked your best and worst nights for three months, you should be able to identify some sense of pattern.

The reason for tracking your best and worst days is so that you know when you need to be extra vigilant in preparing your day-to-night rituals. Routines are not just about how to get into sleep, but more what to do in the hours leading up to sleep, especially when you know it might be a tougher night. Knowing your sleep–wake cycle pattern is key to understanding healthy sleep habits.

As the light levels dim from day to night, your body will naturally begin to prepare for sleep. Your temperature begins to lower and your pineal gland will be getting ready to release melatonin. There are some hormone disrupters that are very common in our daily lives which prohibit these natural hormonal processes. The first of these, famously, is the blue light emitted from screens. This light is specific to electronic equipment such as TVs, tablets and phones. It keeps our brains in day mode, prohibiting the release of melatonin and meaning your body doesn't switch as quickly as it should into sleep. One of the most important and sometimes hardest rituals is to avoid looking at phones or electronic devices for up to three hours before bedtime. This has a twofold effect: not only does it stop the blue light preventing your hormone system preparing for sleep, but it insulates you from any stress triggers that come into your phone from email, news or social media. Even messages from friends or family can have a disturbing effect, after all. The main thing to remember is to reduce stress hormones, such as adrenaline and cortisol, so that your brain goes from alert to rest mode.

The best piece of advice I was given when trying to build this into my day-to-night routine was buying an analogue alarm clock. Setting my morning wake-up at 7 a.m. means I know I don't need my phone between 7 p.m. and 7 a.m. Try this for twenty-one to thirty days and see the impact on your sleep cycle: it is a simple but highly effective solution.

When you have been able to build one basic routine like this 7–7 rule (or 6–6 if you're an early riser), you can then integrate other smaller routines, whether that be a fifteen-minute restful yoga stretch, meditation or just a long bath on the days of the month when you know it will be more difficult to turn off a busy brain and achieve a restful sleep state.

Melissa Hemsley, as a successful and busy food writer who has suffered with insomnia, was also someone who highlighted the importance of routines in getting the proper amount of sleep: 'Sleep is my key to keeping happy and healthy. Sleep disruption became a huge part of my life, due to work and high-adrenaline events or live TV. Historically, if I had big events or live TV, I never slept and I would get more and more worn out. The adrenaline would keep me awake in the week before and then the high from the event would keep me awake after. I have learnt mindful tools to help me manage and prepare for high stress environments or circumstances, they include: 1. not eating too late, so my food is digested before I go to bed; 2. I make sure if I drink alcohol, it's just the one drink and not to use it as a crutch; 3. the temperature pre-bed is key for me – I don't have hot baths at night, as they tend to make me restless and stay awake; 4. no intensive exercise late at night, as it raises my adrenaline before bed. I have a checklist and I run through this, to set me up for the best night's sleep I can, which means I am most effective in these high-adrenaline work moments.'

YOU ARE WHAT YOU EAT

As Melissa points out, one of the other rhythms that can inhibit deep and restful sleep is eating and drinking too late. Your metabolism does work effectively overnight to process your day's food. However, if you trigger your insulin levels by eating any food or sugary drinks, including alcohol, up to two hours before you sleep, it can cause your body to stay alert.

Essentially, you are giving your body energy just before it needs to switch mode.

Therefore, as a healthy habit, try to avoid food and drink up to two hours before sleep. Obviously, this isn't achievable all the time, but if you have been tracking your sleep and you know the nights that might be your worst, which can then in turn be the start of a bad sleep cycle, you can try and focus your pre-sleep fast on those days. If, like me, the fear of going to bed hungry is worse than a bad night's sleep, then drink as much water or herbal tea as possible before bed. Water and teas don't count in this pre-sleep fast. Hydration can in fact be a huge support to slipping into good sleep. Most people shouldn't need to urinate while asleep: your body produces an antidiuretic hormone to counter this. However, as we age, or for those women who suffer with weak bladder function, it is advisable to ensure you do try and avoid excess liquids up to an hour before bed, to help prolong your sleep cycle and prevent being disrupted by midnight toilet trips.

When I was struggling with sleep it was one of the times I was best able to adopt the food and supplement rituals that our Moody nutritionist Lola Ross recommended. Here are some of her top tips for a happy sleep-hormone cycle.

FOODS FOR STRESS SUPPORT

The human body burns through nutrients much faster at times of stress, as it draws on all its resources to cope. This fast burning of nutrients can deprive our neural pathways of the support they need for kicking off our sleep cycle. It is not enough just to treat nightly symptoms of sleeplessness. It is essential also to help manage daily routines with food that can support slipping into sleep.

You can support your body's macro and micro nutrients by drinking two to three litres of water during the day; it is also helpful to consume healthy fats like avocado and nuts as well as amino acids, which can be taken as supplements. Try to avoid stimulants such as sugars, caffeine and alcohol.

FOODS FOR ADRENAL AND THYROID HEALTH

Having healthy adrenal and thyroid glands will provide a huge support to your body in naturally cycling through stress hormones and helping mitigate their potential impact on your sleep. Build an adrenal–thyroid-supporting diet around wholegrains, plant proteins, polyunsaturated plant oils, lean meat, fish and vegetables (limiting raw brassicas which may inhibit thyroid function). Limit refined carbohydrates, saturated fats and stimulants like caffeine, colas and alcohol.

FOODS CONTAINING PROTEIN

Protein provides the raw materials for the brain to manufacture neurotransmitters, including GABA, which is key for sleep regulation. So including protein is essential in any sleep-support routine. The recommended daily amount of protein for women is around 46 grams and can be found in foods other than just the obvious meat and fish: chickpeas and beans, oats, nuts and seeds.

FOODS AND THEIR VITAMIN POWERS

Tryptophan is an amino acid that is the precursor for brain synthesis of serotonin – so an essential component of the sleep cycle. If you are struggling with sleep, try including any of the following foods containing tryptophan in your evening meal: walnuts, eggs, bananas, dairy products, turkey meat or soybean.

FOODS CONTAINING MAGNESIUM

Magnesium supplements are my nightly support. This mineral regulates nerve transmission, which has been shown to soothe anxiety, lighten moods and encourage deeper sleep. Magnesium is depleted during times of stress, which has an effect on the body's ability to relax and access the deep-sleep state that is required for growth-hormone activity. Many everyday foods contain useful amounts of magnesium, including wholegrains, seeds, leafy green vegetables like spinach and fish.

FOODS CONTAINING B VITAMINS

B vitamins are found across the nervous system and levels become rapidly depleted when we're stressed, suggesting the important role they play in helping us manage stress and metabolism, which in turn supports anxiety and reduces risk of insomnia. B6 is a power player in the production of dopamine and serotonin, vital to our happy hormonal flow. B vitamins are found abundantly in the husks of wholegrains, fortified cereals, nuts and seeds, green leafy vegetables, sea vegetables and seafood. For good B6 sources in particular, include plenty of lentils, spinach, cold-water fish and potatoes in your diet.

SUPPLEMENTS FOR SLEEP

Theanine (n-ethyl L-glutamine) is an amino acid found in green tea and has been shown to increase the neurotransmitter GABA which is the ultimate supporting chemical for sedation and getting deep sleep.

Magnesium and B6 is a supplement I take every evening, but I always avoid in the day before 5 p.m., as it can make you drowsy. It is a useful combination for women specifically in supporting relaxation and sleep.

5-HTP (5-hydroxytryptophan) is an amino acid that helps to form the essential happy hormone serotonin. Taken for short periods of time only, it has been shown to enable sleep in those dealing with wakefulness.

Vitamin C – also known as the immunity booster – is essential for adrenal health, allowing the adrenals to respond efficiently to stress; it also helps with energy metabolism during the day, which can lead to better sleep at night.

Estrogen dominance support can be a game-changer for reducing PMS, menstrual pain or heavy bleeding, but it takes a level of commitment and time. It involves a targeted food and lifestyle strategy looking at improving diet, activities and reducing environmental pollutants which place a burden on the liver and estrogen functions. That includes trying to remove any chemical disrupters, often hidden in products such as make-up, which is then absorbed by the skin. It also means reducing stress where possible. However, as I know from personal experience, reducing stress is much easier said than done, so until then, there are certain foods and remedies that may be useful symptom relievers, and encourage good sleep which can be troubled during the luteal and bleed phases of your cycle.

High-fibre foods – found in grains (complex carbohydrates), fruits and vegetables, and legumes – help to promote healthy bowel movements which is one of the ways that the body excretes used estrogen. Including fibrous foods in your diet can also reduce gas and constipation which can be aggravated by imbalanced hormones. I recommend two to three servings a day, particularly during the luteal phase.

Dark green leafy vegetables (DGLV) – such as bok choy, broccoli, green cabbage, watercress, callaloo, chard, spinach and kale – contain important nutrients like folate and B6 that support methylation and glutathione detoxification processes that are involved in the handling and removal of used hormones throughout the cycle. A good aim is two to three servings of DGLV a day. Have them any which way: soups, smoothies, stir fries, raw or lightly cooked.

Water is an essential nutrient, vital for the healthy functioning of our bodies as a whole, but specifically for hormone balance. Water helps with the efficient movement of used estrogen out of the body in both urine and faeces, preventing recirculation that can mediate estrogen dominance. The recommended two litres a day will be beneficial. Use a water-monitoring app if you need help tracking your intake.

Turmeric, used for centuries in Ayurvedic medicine, is a botanical root with anti-inflammatory actions, and has been shown to be an effective pain reliever during bleed-associated cramping.

Magnesium deficiency is thought to be a cause of PMS in some women, so if your levels are low, not only will it help you sleep, it could also reduce PMS. Magnesium malate or citrate can be taken as supplements, but can also be highly effective in bath salts. Having magnesium baths during the fourth luteal and reflect phase of your cycle has for some of the Moody community transformed their PMS and sleep patterns in the month. You can also get a spray form, which can be applied after a shower.

It seems mad to think that the power of sleep has been so overlooked by science until this century: it is the fundamental pillar for our body's ability to operate effectively during the day. It is not a luxury to get sleep or something you should do when you have time or a holiday: it should be a focus for us all as the foundation for a happy and healthy life. Sleep is the cornerstone for your physical survival and, as anyone who has suffered with sleepless nights will know, it is also the cornerstone for your mental health. Your physical and mental bodies live as one, connected by internal hormonal balances and communicating via networks of nerves, glands and organs. Sleep is the nightly reset that allows you to be the best and heathiest version of yourself.

TAKEAWAYS

- Sleep is triggered by the hormone melatonin from the pineal gland. There is a cascade of hormones that are then released throughout the night and your sleep cycle.

- Excessive stress hormones cortisol, adrenaline and nor-adrenaline will prohibit the body's ability to slip into sleep states.

- Sleep-hygiene routines, such as avoiding blue light from a mobile phone up to three hours before sleep, will help your hormones kick in and ensure more-rested sleep.

- Sleep is a time when the body repairs muscle tissues and flushes through excess hormones and toxicity from the day.

LIFE CYCLES: SEX, RELATIONSHIPS AND LOVE

Have you ever questioned why we don't learn about how sex, relationships and love make us feel? At school, we seem to be taught about the function, not the feelings. We learn about what bits go where, which is the least of your concern when you're getting down to it. The bigger questions – How do I feel? What is happening inside me? What do I make of all these emotions? – are left to us to figure out.

Love is a powerful drug, and so is sex. Sometimes they go together and sometimes they don't, but both have important hormones associated with them, and both leave lasting chemical imprints. These imprints can be good or bad, but learning the language of your sex and love hormones can help you unpack some of the more complex and sometimes tangled emotional webs that we find ourselves in.

KEY HORMONES AT PLAY

The major hormones within our sex cycles and life cycles are ones we have already met: DOSE and the ovarian-cycle hormones, estrogen, progesterone and testosterone. These hormones don't just form part of the ovarian cycle for fertility, but also are fundamental to our desire and need for human connection, both sexually and platonically. Hormones connect mothers and babies; they connect us to family, friends and even one-night stands. The hormonal reinforcements we get from these interactions are what make humans dependent on social contact for long-term survival. Isolation and long terms of confinement are proven to cause harm to mental and physical health. Touching another human you love is a form of hormonal medicine.

When I was writing this, the whole world found itself in a situation that made my point in the most dramatic way possible – a global pandemic shut us all inside under lockdown rules, and for many people this meant complete solitude and a lack of human touch for weeks. Technological solutions – video calls and Zoom meetings – kept us connected in some ways, but they could not replicate the feeling of actual human touch. A virtual life is a life without hugs and that is not good for our brains or bodies. And this is because a hug, or a handhold, sets off a specific hormonal reaction inside your body. When we hug someone we love oxytocin is released from the pituitary gland and this has a bonding effect; it also lowers cortisol and blood pressure, making us feel calmer and more connected both to ourselves and the other person.

Loneliness and isolation increase the levels of cortisol and this creates internal inflammation, which can cause our bodies to store extra fats and sugars and make our minds foggy, losing concentration and focus. Your body becomes toxic without the love and touch of other people. There are, of course, two kinds of touch: familial and sexual. In this chapter I will explore the relationship we have to hormones that drive desire, but also how our bodies lust and long for both types of touch, sometimes even at the same time. I will look at familial and platonic touch in the next chapter.

Sex is hormonal and it makes us feel good, but it is also an important function of intimacy. Cortisol and stress can have a huge impact on our libido and desire for sex, as they make us crave more-supporting touch and love rather than lust; in other words, we become driven by the desire for oxytocin as a powerful antidote to cortisol.

After a global pandemic, we can all relate to the longing and need for human interaction as a core pillar in our sense of purpose and comfort, and I know for many there was a huge impact on reduced libido from stress. Even if you were locked up with your partner, there were many people for whom the idea of sex during isolation was simply not appealing at all. I certainly have never longed more for hugs and familiar touch, along with the hit of comforting oxytocin which makes me feel safe, warm and connected to the world.

BONDING HORMONES

The key players involved in love and sex are DOSE. These are the hormones that tell our body that this person or this interaction feels good. They are also what gives us the rush at the start of new relationships and later gives us comfort and cosiness, as we slip into longer-term or more stable patterns with partners. We have met them before, of course, but this is how they relate to your sexual desires and patterns:

Dopamine. Have you ever felt hooked on a person, feeling or even vibrator? Dopamine is why; it's the reward chemical that, when you feel good during or after pleasure, hooks you in and can make you want more and more. It is also one of the stimulants that can lead to addictive patterns of behaviour in relationships and sex – 'chasing' the happy high.

Oxytocin. The naked touch and energy during sex is all part of how you gain access to high levels of oxytocin. This is the hormone that bonds us to another person from a hug, but naked passion means a more powerful connection and release. It can often be the chemical that makes it hard for some people to have casual sex: if you are a tactile person or extrovert who craves others, this hormone can bond you deeply to sexual partners and in some instances 'cloud' your view of a partner in the early stages of a relationship. Bonding with someone through sex is a chemical chain reaction: whether you like it or not, these chemicals will change your emotional state and relationship to that person. Although we release oxytocin every time we have sex, it is particularly strong in the 'honeymoon' stage, as there is often more adrenaline, combined with bonding hormones, from the intensity and excitement of a new partner. This is the hormone that bonds a mother to her baby so it's no surprise that it can contribute to powerful bonds with people when you're naked and intimate.

Serotonin. This hormone is the trigger for oxytocin and stimulates the warm, happy emotions we feel when we meet someone we fancy or are falling in love. Where dopamine keeps you wanting more, serotonin can give you the fuzzy feeling inside. The more we fall for a person, the more our brain and body produce this happy high when we are around them, building a connection between the person and the chemical release. Serotonin is also nature's appetite suppressant. This is one of the reasons why, when you first fall in love, you might stop thinking about food and your brain becomes consumed instead with feeding you thoughts of the other person. Happiness and love make you less hungry and affect ghrelin and leptin, your hunger and appetite hormones.

Endorphins. We get these from exercise and sex, which means you get a double hit when you have good sex since it also counts as a workout. This is another powerful stimulant and highly addictive if you spend too much time chasing the rush and the high. Anyone who has been in long-term relationships can relate to the shift from the intoxicating high of the 'honeymoon' phase to the more cosy and comfortable phase when your endorphins stop rushing every time you touch the other person.

These hormones do not sit in isolation of your wider endocrine system, and therefore the other major hormone cycles, such as metabolism, sleep and ovarian cycles, that we process through each day, week, month and year will also have an impact on them. Hormone disrupters, including stress, environment and food, will affect how our levels of sex and love hormones behave. As with all these cycles, they are connected and make up your unique hormonal code.

It is also important to establish that, when I talk about sex, I mean sex that gives pleasure, and relationships that evoke happy emotions and therefore happy chemicals. There is a brilliant book by sex podcasters the Hotbed Collective called *More Orgasms Please* where they specifically unpack this topic. But, sadly, this is not the experience that all women have of sex. Sex for many women can be part of traumatic past experiences. Shockingly, 1 in 4 women have experienced some degree of domestic abuse, either verbal, physical or sexual, from a partner. Sexual or emotional trauma from relationships builds long-term emotional and physical damage through hormonal patterns of sex being connected to our fight-or-flight responses and cortisol.

Hormones are intrinsically linked to trauma, and in Bessel van der Kolk's book *The Body Keeps the Score*, he outlines the research that directly links domestic and sexual abuse to insulin resistance and subsequent illnesses such as obesity. Stress hormones from trauma can create emotionally responsive eating patterns such as bingeing and this, combined with the increase in cortisol from the trauma itself, leads to insulin resistance and interferes with the appetite hormones ghrelin and leptin. This means the body's natural ability to process fats and sugars is limited and this, in turn, subsequently limits the production of happy hormones. Happy hormones are essential for our body's balance and, when sex or relationships are linked to trauma, women can lose this access point to their feel-good hormone stimulation.

When discussing trauma with many women in the Moody community, there has been a common theme of women's stress and fear of penetration, which can be built from past negative or abusive experiences of sex. Ellamae Fullalove, in talking about her experience of MRKH, shared a particular trauma she experienced from penetrative dilation therapy. Commonly prescribed to women with MRKH, this is a very invasive treatment which involves the internal use of instruments to stretch the vagina. It is prescribed because the vaginas of MRKH sufferers are often shorter, making penetration painful or impossible, even though penetration in sex is not really necessary. For a sixteen-year-old, newly diagnosed with MRKH, this was a hugely stressful process. Ellamae found a way to work with both the physical pain and emotional trauma through masturbation and orgasms as therapy. 'Dilation can be traumatic, it is a hard and sometimes a long process, it is not something that doctors should ever force women to do. There is so much shame associated with it. I actually had my first orgasm during one of my dilation sessions, as I realised it was easier to do if I gave myself pleasure. Yet there was and still is so much shame associated with the idea that pleasure and clitoral stimulation can relax us and it makes the process of dilation easier. I wish someone had given me permission or made me feel more normal about pleasure, as this could have saved me a lot of pain during my dilation therapy.'

Research which seeks to rebuild libido and sexual rehabilitation through masturbation and self-pleasure is a powerful asset for the future of women's health and wellness. Digital apps, such as Ferly, are seeking to make this rehabilitation accessible to women through a new way of thinking and building a sexual relationship with yourself. Sex is not just about another person; it is about connecting with your inner self and the happy hormones that can even help you recover after trauma. Happy hormones have a cognitive healing effect. Pleasure and the release of hormones from orgasm and intimacy is good for your mental and physical health and supports hormonal health for sleep, metabolism and ovarian cycles.

WHAT SEX MEANS FOR YOUR HORMONES

Sex is not just fundamental to procreation, but also, for the majority of the population, is a huge part of life and relationships. The fine line between what is a friend and a lover is often easily denoted by the act of sex and intimacy. When I talk about sex, I talk about pleasure and orgasm, rather than the functional aspects of anatomy. I remove penetration from the definition of sex, as most women don't orgasm from penetrative sex alone. In fact, 9 per cent of women have never experienced climax at all. For women who have experienced climax, the clitoris is a fundamental part of pleasure. I am sure I'm not alone in relating to this and being very frustrated that it was only as recently as 2017 that I learnt that a mere 11 per cent of women climax from penetration alone. Although intuitively I have known this my entire sexual life, it also shocked me that a subject as important to almost all of us as desire and pleasure is not taught in school.

There is a dance that happens inside us when we experience pleasure with someone else. This starts with the release of serotonin and dopamine, which can come just from being in the mood and finding your partner in the same frame of mind. Then there is adrenaline, and some oxytocin from touch or kissing. Every nerve and sense in your body is moving and responding to this physical interaction with another person, an act which is signalling to your body and brain to release hormones that contribute to the building of intimacy. Then comes the hormonal crescendo of an orgasm.

Orgasm, climax or the big O is one of the most powerful and quickest ways to hit an elevated DOSE of happy-hormone high. From a hormonal perspective, an orgasm is an extremely good way to stimulate serotonin and dopamine if you are feeling low. Orgasms reduce cortisol, which as we know is a stress-induced hormone that can play havoc with all your other hormonal cycles. And you don't need someone else to get one. Having orgasms isn't just nice: they help regulate your overall hormonal health and therefore your overall happiness. It's a hormonal meditation state, as this moment clears the mind and fills the body with a combination of all the best sensations and chemicals. For anyone who has experienced one, I don't need to explain any more.

Bonding with another person through orgasm can mean emotional judgements about this person become powerfully interconnected with these mind-altering chemicals. Interestingly, there is a difference between how men and women release oxytocin post-orgasm: women continue to release the hormone up to three hours after the act, while men stop producing it almost immediately after ejaculation. Oxytocin is also nick-named the cuddle hormone and is one of the reasons women often want to spend more time close to their partners after sex.

This means that in heterosexual casual sex, there is an imbal-ance on the hormonal bonding front. Women bond for longer during and post-sex and this can mean it is harder to discon-nect from sexual partners quickly. I was confronted with the cruel reality of this quite late in my own life, when I had my first one-night stand. I didn't have one until I was thirty, due to being in long-term relationships throughout my twenties. I therefore had no idea that you can bond with someone so profoundly after one night, even if it is only a throwaway experience for the other person involved.

This effect is even more heightened in the second half of your cycle, when progesterone is higher and serotonin lower, meaning the craving for human contact and happy hormones is stronger. To make matters more complicated, we have spikes in our testosterone in this second half of the ovarian cycle, which can surge our libido, so the pull to have a one-night stand is perhaps even greater. For couples, on the other hand, this is often a great time to have sex: the second half of your ovarian cycle can create much more connection with your partner, due to the oxytocin, dopamine and serotonin hit your body is craving.

For all the single ladies out there, and I speak from my own experiences here, this second half of the cycle is the time when you can feel more vulnerable and scroll dating apps or look back through your list of previous partners to see if there is anyone who might serve this sexual urge. But beware, as the hormonal imprint can cloud your emotional judgement. It is the fourth phase of my cycle when I am most at risk of making bad partner choices. Hormones can evoke powerful emotional responses and can at times be hard to decode as love, a crush or just too much oxytocin.

This is another reason why I choose not to drink alcohol at this time, as the combination of casual sex, alcohol and hormonal shifts is a recipe for decision-making disasters for me. When you think you might be falling for someone you've been sleeping with, one of the key questions to ask yourself is: Do I feel these emotions more at specific stages of my month? At one point while researching Moody and using my own sex life as a hormone lab, I decided to give up sex entirely and only commit to kissing, but even kissing all night leads to a certain amount of oxytocin release and therefore bonding. I now know, from tracking my lust and longing for someone, it takes me about seven to ten days to cycle through these hormonal imprints, and by day ten I begin to think more clearly about what I really feel for the other person, outside the fog of hormonal sex highs.

DO-IT-YOURSELF

The orgasm as hormonal medicine doesn't need to involve someone else. Masturbation releases your DOSE and even though you are solo, there will still be oxytocin, as bonding with yourself is sometimes as powerful as bonding to a partner. Masturbation is the most effective and quickest route to women accessing the hormonal high of orgasm, as 95 per cent of us orgasm this way. Masturbation increases the blood flow throughout your body, meaning it can also help reduce stress and regulate insulin levels. It is therapy and should be administered as a way to help you relax and feel happy; it's the ultimate self-care. It is also a way for us to discover our own bodies and pleasure triggers, enhancing our overall sexual experience.

The internet is not short of videos on 'how to make women orgasm', but seemingly the research in this area is very much lacking in understanding the power and importance of pleasure for mental health. As I keep highlighting mental and physical health are as one, so we need both nourishment in physical and mental forms. In each chapter I have tried to give a set of food and supplement solutions that can support the hormonal patterns, imbalances or cycles in which we find ourselves. The key to this chapter is not just to understand the hormones at play, but how effective your mind and body connection can be in supporting them. If I could leave you with any one tip for how to optimise your sex life, libido and even pregnancy hormones, it is to tap into DOSE via masturbation. If you don't own a vibrator, my best recommendation is to go and buy one as soon as possible and give yourself some happy hormones via an orgasm. This can have an immediate effect on your mental and physical health, metabolism, sleep and even ovarian cycle. It is stress relief and meditation in one easy and self-induced moment.

THE STAGES OF LIFE

Sex isn't something that happens only during our peak fertility years: humans do have sex for pleasure after all. Therefore, we have sex throughout life and the hormonal life stage we are in, from puberty to post-menopause, will have an impact both on how our bodies respond to partners and who, why and how we desire.

PUBERTY

Puberty is a phase when a complicated combination of hormone surges, sexual awakening and inexperience plays a part in every decision you make. It is a confusing time for most of us, and of course it is also the time when a lot of us begin our sexual lives. You are developing so quickly and your hormones are working overtime to help grow your brain and body. It's a metamorphosis from one body phase to another: we develop boobs, curves and hair everywhere. We also go through skin transformations: famously this is often a time when breakouts plague our everyday. This can get to the point of being totally debilitating due to hormonal acne.

These hormones don't just affect our physical appearance, but who, why and when we are attracted to partners. I grew up in the nineties and it didn't feel like an era of huge gender and sexual fluidity compared with the current decade. My own education about sex consisted mostly of conversations with my mum where I squirmed and recoiled whenever she approached the topic. Instead it was my own life lessons that led me to understand who I fancied, but I certainly never thought about how the hormones inside my body were in some sense finding their feet in the same way as my sexual self.

So what was happening in my body? A girl's ovarian cycle is triggered and puberty begins when gonadotropin-releasing hormone hits the pituitary gland. This causes a domino effect, which then triggers the release of luteinising hormone and follicle-stimulating hormone, bringing on a girl's first bleed. This is when estrogen, progesterone and testosterone begin their monthly cycle through a young woman's body, but alongside growth hormones and ovarian hormones, there are also other new cycles associated with lust, love and relationships.

Together with body and brain changes, it is the time in our life when we begin to form strong bonds outside family. Friendships can be as volatile as romantic relationships, as the hormones associated to puberty impact on our moods and emotions. We can form deep friendships during this time, which can feel polarised by the very high highs and very low lows of the hormonal phase. I am lucky in that my closest friends are still those I met at this stage of my life. We weathered the highs and lows together; we were not always kind to each other during this tumultuous period, but those hormonal bonds interwoven with shared experiences have connected us forever.

Romantic relationships also begin to form during puberty and our early libido and lust is sometimes erratic, driven by more volatile spikes in our testosterone and estrogen activity. It can feel as though you're on an emotional and sexual roller coaster that is going at full speed. The highs feel off the scale but the crashing back down to earth can feel much harder. We all know the caricature of teenage boys who can't control their erections and their hormones playing havoc with their drive to masturbate almost non-stop. Obviously that is a very crude stereotype. However, a version of it does happen for teenage girls too, libido and sex drive meaning the urge to masturbate and lusting for people in your class can be at times confusing and exhausting. The pace at which your attraction and lust for people peaks can be from day-to-day and this is what makes teenage desire so confusing. Which I am sure everyone can relate to, as this high–low phase is one we all go through.

I look back at some of my early sense of lust and love and realise how much I was guessing at what my body was doing, rather than being informed. I was so hard on myself and expected to fall in love like they do in movies. My friends and I call this Disney Princess syndrome, which is the idea that we have been conditioned by these notions of being 'swept off our feet' by a Prince Charming. The high-school relationships I had certainly fell short of this; not surprisingly, as both sexes at that age are just testing out their bodies, boundaries and sexuality, while the heart–mind–sex connection is still under development. Hormones are not taken into account in any romantic narrative I have ever seen.

In fact, what happens during this phase of puberty is that we're building our hormonal pathways and experiences, helping formulate our understanding of what feels good and bad.

These hormonal pathways are what help us establish good relationships in the future. Bad experiences can bring trauma that can have negative ramifications such as issues with appetite and food, which can lead to eating disorders. Healing from this trauma can involve both physical and mental therapies, and as we have seen there is more and more research going into sexual rehabilitation treatments, which can involve masturbation as a way of retraining your body to respond to pleasure.

The issue I had when I was an adolescent was that I wasn't given permission to test any boundaries or listen to my body. There is almost a dictionary of words used to disarm and undermine women's sexuality and 'slut shame'. There is even a cultural dismissal of female masturbation and pleasure, with most pleasure narratives from porn to mainstream media being driven by the male gaze. In the nineties there was certainly no *American Pie* movie for girls, where women's awkward masturbation stories were just part of the cultural norm.

So, for me, working out who and what fitted my hormonal code and evoked happy hormones or stress from desire, libido and sex was a minefield. For most young women I speak to, it seems that still is the case, even if the cultural landscape is beginning to change with films like *Booksmart* showing there is more to being a teenage girl and boy than just the male perspective. If I could have one wish for the next generation, it's that the understanding of hormones means everyone is simply less hard on themselves and feels less guilt about the natural cycles they are going through. Some days it's OK just to say to yourself, 'I am too hormonal and I am going to be kind to myself today.'

Puberty can be an especially challenging time for some. Kenny Ethan Jones is a trans man looking to revolutionise the way the world sees gender. He made history by being the first person to front a menstrual-care campaign as a menstruating man. He is an activist not just for the trans community, but for a world needing to be re-educated about unheard and misrepresented voices. He describes how the hormones that hit him at puberty affected him: 'Puberty for me was a hugely challenging time. The hormones and estrogen that were changing in my body at this stage made me a very unhappy person. I would get huge mood swings as I started to grow boobs and menstruate. I just didn't feel like me. As soon as I started taking testosterone the mood swings levelled and I began to click into the person I was meant to be, I got back on track and found happiness again.' He remains curious about why his body reacted so badly to the hormones of adolescence: 'The question I want to ask about hormones is why did the hormones I experienced in puberty make me feel so removed from my person I was? I became a different person when puberty began and thankfully I was able to identify early this was due to my hormones and changing body.' The difference once he was finally able to take testosterone – what he calls a 'second chance' – was like night and day. Testosterone – now in the form of Testogel – made this second adolescence an entirely different experience as he was finally able to allow his body to develop in the way that felt right to him.

As a trans woman, Hannah Winkler echoes this viewpoint: 'You go through a second adolescence, which means you need to be prepared for a turbulent emotional and physical hormone journey. What I noticed in my transition was that, when I began taking estrogen, I came out of the depression brought on by gender dysphoria. Once the rocky first phase, which lasted roughly six months was over, there was a sense of euphoria that set in, as the depression began to lift.'

Puberty is a challenging time for everyone and hormones are at the core of this turbulence. We all go through a hormonal metamorphosis during this time. What we all know of puberty from experiencing it is that you come through the other side; and you will come through it with new-found skills to navigate life, friends and family.

SETTLING IN

Being free and experimental with learning about your body, what feels good and what doesn't, is not to be determined just in puberty. It is a lifelong pursuit and will change at different life stages. Your hormones will alter as your body moves from puberty into new stages of your life cycle and therefore so will your lust, desire and attraction to partners. This is not to say I have any opinion on whether we should have one or many partners in our life, it is more a simple view that hormones play a part in how you feel about your partners, dependent on how old you are, your life experience, stress and all the other hormonal factors we have already talked about. Our sex hormones are as much part of our evolving and changing hormonal self as our sleep and metabolism cycles.

After five to seven years of early ovarian cycles, your pattern tends to emerge. There is a sense and intuition that increases as you settle into your body. In healthy women, this is typically your most hormonally balanced time. On average, women from twenty-five onward begin establishing their hormonal ovarian cycles and technically this is the age when women move from adolescence into a new chapter.

Your hormones evolve and so does your sexual self. It isn't just about hormones: by twenty-five you have also had more life experience and you begin to learn both mentally and physically what feels good. We tend to feel more confident in life and the lessons we have learnt to date, and this feeds into our sexual self. However, society and stress are often more to blame for challenges in our sexual freedom. Stress that can occur from feelings of inadequacy or body confidence can have a hugely detrimental effect on our ability to feel free or even safe in our sexual development. In my own early experiences of sex and relationships, I would often recoil with embarrassment at how I had emotionally or physically 'behaved'. Now I realise that my brain and body were going through a change. At that stage I didn't know enough about my patterns to understand that there was going to be a certain amount of unknown and uncertain hormonal development. What I know now is everything I went through in puberty and my twenties was normal for me and I wish I hadn't been so hard on myself.

MENOPAUSE

The next major hormonal milestone for many women is the most feared, and certainly for many women is a topic rarely discussed until their group begins to hit the cycle. As a woman in my thirties I have been shocked at how little we are taught about a life cycle that every woman will go through.

When researching hormones, what I have found most fascinating is just how much confusion and fear is associated with normal hormonal changes that happen to most of us. The fear is perpetuated through a lack of clear and simple information and is particularly strong when it comes to the menopause. Although there is now a fantastic mix of books available, such as *The M Word* by Dr Philippa Kaye and *The Second Half of Your Life* by Jill Shaw Ruddock, I want to share my own very simple 101 on the hormones, phases and untold benefits often missed in this important life stage.

There are three phases.

Perimenopause. This phase often begins eight to ten years before menopause occurs. It is a time when your estrogen drops overall and you often experience irregular periods. You may begin to experience some of the symptoms such as mood swings and temperature fluctuations commonly attributed to the phase. What women tend to notice is changes in their ovarian cycle, alongside the general sense of emotional unrest, shorter temper with partners, friends and colleagues, sleepless nights, hot sweats in the day and at night. These are all worth tracking and tuning in to, so you can begin to see patterns, which you will then be better able to manage.

Menopause. The switch is marked and confirmed when you have not had a period for twelve months, which often happens in your late forties or fifties. From this time onwards your body goes from producing ovarian hormones each month, and cycling through from high estrogen to lower, into a steadier level of progesterone and estrogen across your months. The moment of menopause having occurred is no more bleed. This marks the end of your fertility cycle, but it is also a new beginning.

Post-menopause. This new hormonal chapter brings other cycle shifts as well. Your metabolism and sleep cycles change, often sleep becoming deeper and metabolism slower. Some women gain weight and some experience changes in hair growth, from thinning head hair to new facial and body hair. This is a time to reconnect with your new body and to forge new routines for social life and health life. There is a shift from a cycle of health focused around a monthly cycle to a focus on how to optimise your metabolism- and sleep-cycle hormones on a quarterly basis, as a way of finding a new timeline for your body and mind.

The phases of menopause that are most important to track are the first and second, as this is when your body is switching. This is when moods and symptoms can be an indicator of how far into the phase you are. Tracking will allow you to monitor and therefore support some of the adverse effects. Most women will have been managing their cycles on a monthly basis with PMS, so changes in hormones and their effects are not new news. But the changes can be less regular and this is why technology can serve a powerful purpose in helping identify if there are patterns to get you ahead of the challenges.

The first thing to track during perimenopause is your period and its irregularity. The longer the time between periods tends to mean the further into your menopause phase you are. There are other signals and signs that indicate the body's hormonal changes. These include more prominent and frequent changes in temperature, mood and sometimes symptoms such as weight gain, facial hair growth, or hair, skin and nails becoming more brittle and thin. All of these are worth tracking, as they are also linked to the lower levels of estrogen. Sex drive can be hugely affected as the body moves from one life stage to the next, and vaginal dryness is also linked to the lower estrogen. Everyone has different scales and levels of these symptoms and moods, in the same way as every woman has a slightly different physical pattern to her monthly ovarian cycle.

There are brilliant ways to support these changes with natural supplementation, in particular:

Vitamin D. This can help with serotonin production, as this hormone becomes reduced through menopause and as we age. It can be taken as a supplement, but is also found in foods such as fatty fish, cheese and egg yolks. Vitamin D is synthesised in receptors in our skin via sunlight, so the more sunshine on your skin during this stage of life the better, but ensuring you are using high factor sun lotion or block to protect your delicate skin from UV.

Vitamin E. This is also available as a supplement and can be found in food such as almonds, hazelnuts, avocado, broccoli, sunflower seeds and squash.

Prebiotic and probiotic. This is a combination supplement that supports the healthy functioning of your existing gut microbiome. Changes in your estrogen hormone levels at this time can affect your metabolism and gut health.

There are also excellent combination supplements on the market to support the various symptoms and changes to women's bodies during this phase.

Other ways women choose to manage this stage is to go on to a course of hormone replacement therapy (HRT). There are two kinds of HRT available: one bioidentical, the other pharma-ceutical. Bioidentical HRT is derived from plant-based estrogen compounds, from plants such as soya, while pharmaceutical HRT, much like the contraceptive pill, is made from a synthetic hormone compound. It is best to discuss the options with your doctor, but be sure you track your moods and symptoms ahead of the appointment. Much like the contraceptive pill, HRT has been an effective support for some women, with others finding it causes issues and side effects. Making an informed decision around what medication to take, based on how you feel before and after taking it, can be a good way to manage your peri-menopause and HRT if you decide on this route.

The biggest taboo and emotional shift that is often overlooked, when talking about menopause, is that it marks the end of your fertility cycle. This is perhaps partly due to a complex social system that sadly has linked women's worth and social standing to their fertility. It is problematic for many reasons, not least for women who struggle with conceiving. The emotional impacts of feeling less than, or being ostracised from society altogether due to hormones they simply cannot control, can lead to debilitating mental illness and depression. Hormones are as much genetic potluck as what colour hair and eyes you have. Ultimately your hormones are a part of your internal make-up.

Menopause can to some women seem so final, but it is also an identity shift. It is strange to think this cycle is surrounded by negative connotations, when actually women being released from their fertility cycle can evoke whole new life motivation and energy, not to mention the fact that we no longer have to worry about bleeding or contraception. Because of the way society views our fertility, it does take many women time to adjust to a new chapter. Feeling low or feeling disconnected from your past self is a process that we all go through at different stages of life, but it is made more prominent by the biological shift at menopause.

But it doesn't have to feel like this, and doesn't for every woman. Clare Gorham is someone who has gone through the menopause, and certainly doesn't feel any less a woman for it: 'As a "menstruating woman", your sense of being a woman is largely determined by how severe or compromising your cycle is: for one week, you are weepy. For another, you are raging and irrational. The third to fourth, you are really horny. Then the final fourth to fifth week you have a dull/extreme ache, feeling like your womb is falling out of your rectum. So no, in short, losing my cycle doesn't make me feel less of a woman, or that it's negatively changed my sense of womanhood. Quite the opposite. I'm no longer governed by a mercurial, hormonal, self-sabotaging minx sitting on my shoulder, whose pendulum-like moods sometimes/often knew no bounds. Now I feel calm, consistent, measured, more dignified, less irrational. I feel anchored for the first time in my adult life. I'm not a bowl of hormonal minestrone soup. I'm now a delicious, well-seasoned gazpacho in the summer or warm Heinz tomato in the winter. My favourite home comfort and nourishment.'

At Moody we view menopause as a hormonal rebirth and I am adamant that before long we will all be having cake to celebrate our menopause second birthdays, using this happy moment as an excuse for a large injection of DOSE. Menopause can be a time to embrace, a new life chapter with new goals for how we want to live without the hassle of bleeding each month.

SEX AND STRESS

As I have illustrated throughout, estrogen has a profound effect on our bodies when it interacts with all the other hormones in our system. For women there is an important interplay between estrogen and testosterone that occurs in connection to libido, but this can be suppressed by elevated cortisol and stress.

Sex and the hormones released during and post-pleasure fluctuate across our monthly ovarian cycles and our hormonal life cycles. The stress hormone cortisol in particular can be a huge factor in women's loss of libido, and can affect not just the mental desire, but also the hormones that control the vaginal discharge which makes for our natural lubrication. In other words, stress puts both a mental and physical blocker on pleasure seeking. As almost all women know, lubrication is a core part of good sex. This lubrication, as we have seen, fluctuates throughout a woman's monthly cycle and discharge changes in colour, consistency and quantity, as it is connected to estrogen production. The discharge therefore can signify changing hormonal levels. Low estrogen and very low progesterone can disrupt HPA-axis pathways and trigger anxiety. This can then cause more uncomfortable sex, which in turn creates cortisol and stress. In other words, you begin to build a cycle of pain and frustration associated to sex and this can lead to emotional and physical blockers. Dr Sarah Welsh, one of the founders of HANX condoms who we met earlier, believes: 'There's not enough education about why lube use is positive in sex, and it's perpetuated by cultural messages saying that we shouldn't need to "have" to use a lubricant.' A survey HANX undertook of 5,140 people found that 76 per cent of respondents associated wetness with arousal. 'More so,' continues Sarah, '59 per cent of the women surveyed said they regularly experience discomfort during sex, while 95 per cent said they experienced it at some point. This is something I encountered countless times while working in sexual health clinics, and what we want to help overcome. By opening up the conversation around sexual wellness and preconceptions and having frank and honest discussions about them, we can help banish the taboos and stigma.' She and her colleagues have therefore developed a natural-as-possible lubricant, designed to have a pH of 4.5, similar to that of the vagina. The lube is water-based, meaning it is compatible with latex condoms, unlike oil-based lubricants which dissolve latex and break the condom. As Sarah puts it: 'We are non-scented and as near to the real thing as possible!'

A need for lubrication can be particularly true after menopause. One of the most common symptoms that forces this change is vaginal dryness. This condition – also known as atrophic vaginitis or vaginal atrophy – can make sex uncomfortable and sometimes even painful.

Although good sex can be the perfect antidote to the emotional changes and fluctuations that come with menopause, it is hard to find new routines and physical play that work with the body's changing balances. As Dr Welsh explains, 'After menopause, the vaginal tissue becomes more dry and fragile, meaning women are more susceptible to injury or tearing, especially during intercourse. Hence, the need for lubricant is even more essential!' Speaking to women in the Moody community who use the app to track their menopause, we have found many who say it is a time to try new things and reassess what pleasure means. If you can lean in to the changes, you can find a whole new way to access pleasure, from non-penetrative sex to play.

Clare Gorham had been attuned to the realities of what the menopause might do to her sex life, having heard various horror stories from family and friends. But five years in, as she puts it, 'I'm glad to report that I haven't had horrendous dryness. My Little Lady is still lubricating herself rather well. Thankfully!' When it comes to her libido, she adds: 'My libido has lessened – but it's still definitely there. I'm even more grateful for those inimitable feelings of sexual arousal – because I know other menopausal friends can't feel it in that same way. I know it might rescind at any moment, so I really appreciate it. I think with age, and an increased emotional maturity, comes an increased sexual confidence anyway. I feel more attractive now than I did in my thirties – even though I know I'm about two stone heavier with receding gums and a matching hairline ... So I guess I'm fortunate; I'm not a dried-up, atrophic, orgasm-free zone. Yet!'

EMOTIONAL MILESTONES

Whatever stage of life you're at, we all well know that we can be subject to heartache, heartbreak and crushing crushes. What we may not be so clear on is that these moments themselves have significant effects on our hormones.

When we go through grief or loss, the first reaction of our bodies is usually to give us, in a short burst, both adrenaline and cortisol. This is your body's response to trauma, to try and push you through.

Next comes the 'wearing-off' stage, where the immediate energy from the stress hormones disperses and the sadness or longing for the other person kicks in. At this point we begin to crave the happy hormones that we have connected to the lost other. One of the hardest hormones to flush out after a break-up or loss is oxytocin. The length of time you were with your partner can determine how long it takes to reset the mental and physical pathways of connection. The good news is that we don't run out of oxytocin: in fact, studies show we develop more of this power hormone as we age, meaning the older we get the more we can bond and rebond with new partners.

It is while we are waiting to flush these hormones out of our system that we should try and utilise our treasure chest of solutions to evoke the happy hormones we crave. Although it won't be possible to recreate the love or lust we enjoyed with our former partner, we can assuage our hormones with comfort. Unfortunately, we tend not to make the wisest choices at this point. This is where, in Hollywood movies, you see the heroine reaching for a tub of ice cream; when I have gone through heartbreak myself, I have historically indulged in cake, chocolate and any sweet treat I could get my hands on. Although this served a short-term need by giving a quick burst of energy and happiness, it also meant the sugar-high combined with the lower energy and low serotonin was a heartbreak accelerator.

There are better ways. Self-care that involves friends, family and other loved ones is a much more hormonally balanced cure for heartbreak hormones. What is very good to know is, as with all hormonal cycles, the old hormones are flushed out and new ones develop. It takes time to flush away the emotional and hormonal imprint of another person in your life. The truth is, your body will regenerate, as new happy hormones are created and new bonding hormones develop. One day you will wake up and the memory of your past partner will no longer reduce you to tears. One day they simply won't even appear in your mind. The more happy hormonal processes such as seeing friends that make you laugh, taking long baths (if, like me, this is your happy place) or walks in nature, will defuse the downturn and low moods with self-induced happy high's direct from your brain to body. What are your happy places and processes? This is important for you to know, as these will be the hormonal medicine you need to flush out the pain. Employ them as regularly as you can, because the more you do them, the more you will help your body reset after heartbreak. These hard-earned skills will serve you well in the future, as heartbreak can happen on smaller levels almost every month, given that life is full of highs and lows. When you know what your happy list is you can employ it at other times, such as not getting a promotion or missing out on a life goal you had your heart set on. In the same way, your darkest moments can inspire the solutions that help navigate heartbreak on all levels. Your happy list is a hormonal heartbreak medicine, with its direct access to the DOSE chemicals that will make healing possible. The day you wake up and don't think about that person is the day you know your hormones are ready to get you back out there.

PREGNANCY HORMONES

It would be impossible to write a book about how hormones can become superpowers without talking about pregnancy and post-partum, another set of cycles that women have been intuitively navigating since the birth of our species.

There are of course nine months of hormonal changes that happen in a woman's pregnancy cycle, but the hormones don't end there: they continue right through until you stop producing milk or your ovarian cycle resets, which can for some women be up to a year after birth.

Women's pregnancy hormones are categorised into three 'trimesters', but I would argue there are more like seven 'mesters', if you take into account the hormonal phases that occur after the baby is born and the emotional shifts women have to adjust to, from birth to going back to work or changing home–work cycles to adapt for a new life phase.

The hormonal element of the emotional stages we go through during pregnancy also, of course, feed into the gruelling pain suffered by women who go through miscarriage. We often ignore the difficulties women face when having to process and cycle through these hormones, even if their pregnancy ends earlier than birth. Not only are these women dealing with grief, loss and heartbreak, but also the leftover hormones in their systems. It takes time for your body and mind to process such a profound event. Therefore, I feel it is important to share some of the science behind how hormones make you feel during the trimesters of pregnancy and post-partum. It is by far the most fast-paced and extreme of hormonal changes that women may experience.

As with ovarian, metabolic and sleep cycles, pregnancy-hormone cycles have an impact on daily, weekly and monthly moods and symptoms, all working in unison to create a new state of normal for your body with its plus-one. All the glands in the body's endocrine system slightly reconfigure their levels and cycles to adapt. As with all cycles, everyone's code and levels are slightly different and unique, but there are some average fluctuations, which can be tracked, to see how you are emotionally and physically navigating your way through this hormonal hotbed.

PHASE ONE – FIRST TRIMESTER: MONTHS ONE TO THREE (ONE TO FOURTEEN WEEKS)

It all starts with the elevation of three signals: estrogen, progesterone and a hormone almost exclusively saved for pregnancy called human chorionic gonadotropin (hCG).

Progesterone. This is the preparation hormone for getting pregnant: it rises to allow for the fertilised egg to implant within the uterus lining. It also signals to the muscles in your uterus to relax, in order to expand and grow alongside the baby. As I have outlined, progesterone is a 'sedating' hormone. It not only acts as a muscle relaxant internally, but can cause the body to be more sensitive to pain, smells and your wider environment. It is the hormone linked to women's often heightened sensory awareness during pregnancy. It can also be the hormone trigger for mood swings, so if you are reduced to tears by a weepy movie, or find yourself craving exotic foodstuffs during pregnancy, this is the hormone at play.

Estrogen. In the first trimester estrogen is released to help regulate the increased progesterone levels, while also stimulating new growth of the endometrial lining, which is essential for the baby's development. As your power hormone, it is also the hormone that promotes blood flow and gives that 'pregnancy' glow that women talk about in skin and hair. In the first trimester, as hormones are setting themselves up for this new structure, it can sometimes make women feel restless. And with higher blood flow it can put additional pressure on organs such as the kidneys and liver. This means women can often need more hydration, while sometimes also experiencing an almost unending need to use the bathroom. This trimester is one of flux, and the emotional shifts that occur in unison with the physical changes can become more and more noticeable as the phase develops.

hCG. This is widely thought to be the hormone connected to the nausea or morning sickness most commonly felt by women in the first and sometimes second trimesters. Its chemical function is to grow the placenta, which will be the nutrient source for the baby throughout the pregnancy. This hormone elevates rapidly in the first trimester, doubling in levels almost every few days. It often hits its peak at eight to eleven weeks, which is when some of the more uncomfortable early symptoms such as morning sickness tend to peak as well. For some women, this is when those sickness symptoms begin to subside, but as with all our hormonal codes, other women continue to have higher hCG throughout and this is often linked to more sustained periods of sickness.

Towards the end of the first trimester, at around week eight, women can have an elevated or spike in libido. This has been linked to the shifts in hCG and of course a reduction in feeling sick, and also comes with changes in discharge and vaginal lubrication. Estrogen has elevated the blood flow, meaning heightened sensory awareness, which many women report affects clitoral sensitivity, making the happy hormones from orgasm even easier to access. Sex as pleasure at the end of the first and into the second trimester can be extra fun and extra happy, as the hormones released alongside the changing balances in your body can turbo your orgasm high. This can also act as a great antidote if you're struggling with more pronounced symptoms and mood swings.

PHASE TWO – SECOND TRIMESTER: MONTHS FOUR TO SIX (FIFTEEN TO TWENTY-SEVEN WEEKS)

This is often known as the best trimester. It is for a lot of women where they begin to settle into their body's new hormonal levels. It is the time where women relax into their pregnancy and the new body and mind they are living in, but also where they tune in to some of the less comfortable aspects of this new normal. The big shift in hormones affects water retention, meaning this is often when swelling across the body can occur, with aching legs and lower body cramps, congestion and even haemorrhoids and varicose veins. All these changes can be tracked, to see how they shift from phase to phase and to know that they will pass, as different levels within your body regulate and elevate.

As estrogen levels continue to rise, nails and hair may become thicker, longer and glossier, contributing to the glowing state women can have in this phase. This is also where you start to notice your body changes: boobs will be getting steadily bigger day by day and of course your belly grows, which can have knock-on effects to skin elasticity and stretch marks.

During the second trimester there are some new hormones that emerge:

Melanocyte-stimulating hormone. This affects skin cells and colour. It stimulates changes in skin pigment around your belly and nipples, and can also make freckles and moles become more pronounced. It may even out the skin tone, meaning women often report feeling as though they look great and are comfortable in their skin.

Cortisol. Although elevated cortisol is a sign of heightened stress, it has also been linked to a more alert sensitivity for women during pregnancy. The hormone rises during the second trimester and is actually key in supporting your metabolic rate and glucose blood sugar levels. Insulin and blood sugar levels during pregnancy are often linked to cravings and it is important to keep an eye on them.

Human placental lactogen (HPL). This is released from the placenta and stimulates the breasts for lactation. It appears to regulate metabolism of fat and carbohydrates, and stimulates insulin, balancing glucose in the mother and foetus. It can also be linked to insulin resistance during pregnancy, which in some women can lead to gestational diabetes in the second trimester. Listening to your body's cravings, and tracking your moods and symptom fluctuations, can help identify this early. It's important to keep an eye on what you eat at this time, as maintaining a healthy weight is crucial in helping to manage insulin resistance.

If this trimester has been kind to you and given you some of the extra-sexy hormone boosters that some women report, this is a great time to use the confidence to indulge in sex. It can provide additional bonding and oxytocin connection with your partner, who often during these phases can feel left out. Sex is a great way to share in body-to-body hormone bonding and evoke some extra DOSE.

PHASE THREE – THIRD TRIMESTER: MONTHS SEVEN TO NINE (TWENTY-EIGHT TO THIRTY-EIGHT WEEKS)

This is the final phase and when your hormones shift from elevating rates to preparation states for birth. This is also the phase when the body goes through the most profound physical changes, not just as your belly expands quicker and quicker each day, but also when some of the more pressured symptoms can kick in due to the shifts in progesterone, causing discomforts such as heartburn, indigestion and acid reflux. Pressure and contraction of the lower abdominals can also lead to constipation during this final phase.

It is also a phase when women often struggle with getting to sleep. Sleep at this time can be a huge physical and emotional support, and it is also when dreams can become even more vivid. Some women find it comforting to capture and record their dreams during this phase, as a way of looking back at how their subconscious mind was playing out at night.

Estrogen and progesterone both tend to peak at about week thirty-two. Their levels at this point can be six times higher than your body's pre-pregnancy levels. These hormones at peak are the signal to your body to begin preparing for birth. The body begins magically to ramp up other hormones that will be essential for post-pregnancy.

The most notable of these is prolactin, the hormone that stimulates the breast tissue ready for lactation. It is elevated throughout pregnancy, but in the third trimester it hits a turbo shift and waits for the key signal of estrogen and progesterone dropping post-birth for the milk to be released.

Scientists still don't know the exact combination of hormones and functions that triggers labour. However, what is known is that a rise in oxytocin and drop in progesterone is one key domino in the hormonal sequence that leads to birth.

Oxytocin combined with estrogen stimulates the release of prostaglandins, which soften the cervix to prepare for that final push. This hormone doesn't just bond us, it also helps signal your body through the final phase of birth and is pivotal in triggering the contractions until after the placenta is released. As we have seen, oxytocin is one of the most powerful hormones released to bond a baby with its mother after the pain of childbirth.

The placenta is full of hormones and is literally our body's 'life blood'. It is often seen as very hippy or woo-woo to talk about a big sack of blood and tissue that comes out of women after birth, but if you think about it logically this is essentially the body's food manufactured for human life, which is quite literally life blood. The placenta releases its own hormones during pregnancy that support women's metabolism and insulin levels. It is also known that, when consumed in a placenta smoothie straight after birth or taken over time as dried and encapsulated supplements, it can promote incredible results, not just in supporting moods and hormonal highs and lows post-birth, but also healing the body, skin and hair and even helping in weight management. There is still a huge amount of research needed into women's health and one big area is more understanding about how the placenta can provide other post-partum health benefits to mothers and their babies. It's definitely not regarded as a sexy area of science, but one that qualitatively and anecdotally has proven for centuries to be beneficial for overall hormonal health.

POST-BIRTH

What is often missed for women in the pregnancy cycle is the post-pregnancy hormonal phase. This is commonly called the 'fourth trimester' and represents another new normal for your body, as your skin, muscle and tissue begin to change again post-birth.

There are some particular hormonal changes that happen at this stage, which are worth tracking. Oxytocin is released as women begin to breastfeed, and it continues to bond mother and baby. Low oxytocin can occur for many reasons and is the hormonal link between women who can be at risk or experience post-natal depression (PND). Low levels can paradoxically also be triggered by breastfeeding, as this can accelerate metabolism due to the amount of energy your body is using to produce and express milk: this speed-up can have a knock-on effect on your thyroid and TH levels, which then in turn can have an effect on your other glands and the production not just of oxytocin, but also of the other happy hormones serotonin and dopamine. Women who struggle or can't breastfeed can also suffer due to the body not producing oxytocin during the feeding process. The reality of PND is all women could be at risk of it, and the way to spot it is tracking your moods and symptoms post-birth, specifically looking out for signs that your oxytocin is low. The aim is to keep an eye not just on how your baby is developing, but on how you feel and how you are connecting to your baby.

Karla Vitrone, one of my co-founders at Moody, was driven to join my quest for hormonal education after her own experience with PND. She describes with great clarity how it made her feel and why she decided to dedicate her career to bringing more awareness of the condition:

> After I had my first baby four years ago it came into high definition how little I (and most women) know about their bodies and the chemicals that control us. I suffered from post-partum thyroiditis post-child-birth, which triggered Hashimoto's that had been lying dormant. My TSH was over 100 when they discovered it. My doctor delivered the news shocked that I was still standing and operating. I was lucky enough to have a private obstetrician in New York who took blood tests around six months post-partum – that is not standard in the UK or US. A lot of stuff then started to make sense: the post-partum anxiety, the hair loss, the excruciating bowel issues and the memory loss. It was the latter that really scared my husband: I felt scattered, low and I had lost my sharpness. I essentially had lost my joy and my personality.
>
> As I started seeking help and learning about how your thyroid controls pretty much everything that makes you, well, *you* – as well as the relationship between your hormones and neurotransmitters (my serotonin had sunk scarily low) – I grew even more concerned not just for myself, but for all women. We are taught to suck it up, to deal with it. Who feels great anyways? Well actually that is exactly what you deserve.
>
> The fact that this affects 1 in 10 women is why we all need to talk and listen more to our hormones – to learn and take control. When you know your baselines, you trust better to flag and alert people to when these emotional and physical baselines are off. My road to recovery was long, hard and expensive – I believe in a future world, where women have the tools and knowledge to help avoid long-term damage that can be done from these debilitating hormonal experiences.

The challenge with PND, like most hormonal conditions affecting women, is that it is shrouded in social stigma. Sufferers often say they feel a great sense of shame and fear, alongside the loneliness and despair of struggling to bond with their baby. It can even lead to cases of manslaughter and suicide, which is why the lack of research into female health conditions such as this is a gross societal injustice. Women should and will continue to share experiences and those who have suffered or are suffering from this crippling illness are often the pioneers driving the agenda forward for the future of all human health. PND doesn't just affect mothers, it also affects partners and babies: research into this area is of collective importance for human happiness.

What seems true not just of PND but of many things relating to women's health is that society has got the narrative wrong. This also goes for orgasms, placentas, sex and menopause. The focus should be not about women sorting themselves out, but about a collective support network. Almost every aspect of the female sexual and reproductive hormonal cycles needs 'rebranding'. We should be using human connection to build more happy hormones and help each other better understand ourselves. Society focuses on the physical and functional aspects, the 'how to', and misses all the emotional joys and the signals we should be listening out for to help us connect with these cycles for our long-term mental and physical health.

Thinking about how these hormonal experiences and stages might make us all feel can help change how we view not just the process, but also for some of us our past. We should be letting go of some of the guilt that can come from our hormonal life cycles and begin to think of our bodies as instruments to enjoy and listen to. Good sex makes people happy; bad sex can be destructive and make people sad. It's as simple as that. 'What women want' seems to have been a question raised by men, but as women we should also be asking why and what we want, when we want it and from whom.

Hormones have a huge part to play in every aspect of life, but sex for pleasure or reproduction is formative in our very development of self, from the moment we begin to desire. There is no one size for what these phases and stages might be, but we can begin to unlock their power by listening to our own lust, desire and libido patterns. Not to make us better at sex, but to make us happier and more in tune with ourselves.

TAKEAWAYS

- Hormones involved in sex bond you chemically with your partner. Specifically, oxytocin, which is the hormone that makes it hard to break away from sexual bonds.

- Orgasms from masturbation are the quickest and easiest way to release DOSE.

- 8 out of 10 women don't climax from penetrative sex alone. The clitoris is essential for the majority of female pleasure.

- Our hormone levels change across our life cycles, from puberty to post-menopause. This means our libido and desire also fluctuate across our life cycles.

- Sexual happiness is part of hormonal harmony and good for mental and physical well-being.

SOCIAL HORMONES

As I was writing this chapter, much of the world was in lockdown. People were either confined on their own or with small groups of family or flatmates. This had an unprecedented social impact, but also a hormonal one. Humans are social animals. Something we all seemed to learn from lockdowns was that we need other people. Our bodies don't work well without others. Our social networks are critical not just for how we feel, but to inspire us and help release happy hormones for focus, motivation and energy. The chemical happiness certain people and groups elicit within us is one of life's great joys. Whether introvert or extrovert, we still achieve a happy high from meeting friends, family and even colleagues. Our brains release the DOSE which can also make our overall cognitive function sharper and more engaged; this is then paired with the bonding hormone oxytocin, which comes from a hug or comforting handhold and being near people you love. This powerful internal chemical release is often accompanied by other social stimulants from food, wine and even coffee. Obviously these stimulants if taken to excess can cause issues, but there is also lots of science to show that in small amounts, paired with social interactions, they can have a positive elevating effect on our brains and bodies. Coming back to Csikszentmihalyi's theory of flow, these interactions are part of our hormonal life flow. They are how we build social happiness and simple pleasures into our lives, to access a more focused and more consistently elevated experience of daily existence.

The acceleration of the scientific Enlightenment in the seventeenth century has been directly linked to the arrival of coffee houses. This was the first time that academics and thought leaders had convenient places to meet outside their dusty libraries and laboratories: social space to elaborate ideas and muse with friends and colleagues. Somehow, combining company with the stimulating effect of coffee brought a social revolution that sparked a scientific one. Arguably this process of social interactions and DOSE chemicals inspired a new motivation and thought leadership that has defined our scientific world today.

Meeting a friend for a tea or coffee was one of the things I missed most about lockdown during the pandemic. This daylight social contact is not just good for your brain; it is an important part of how we gain social DOSE and thus access our best and most consistent mental states. Only socialising at night or over digital calls, which became the norm during lockdown, does provide some social hormonal stimulus, but it can never deliver the power of daylight days of DOSE we get from catching up with friends in parks, cafés or on beaches.

Almost exclusively, our fondest memories will be social interactions. Think about your happiest memory and it is likely to be with people you love and in a place that you cherish: the memory alone evokes warm, comfortable chemical reactions inside us. You get a rush of DOSE just from remembering a happy time – which is also why, when you feel low, logging back to happy memories can be a powerful antidote. So, other people make our minds and bodies richer, not just in the moments themselves, but even in the memories these moments build within our minds. Recollecting those places and times brings back some of those happy hormones and is medicine you can unlock right now.

These interactions are as much part of our hormonal happiness as exercise, diet and supplementation. There is an elaborate choreography at play here. Dr Minisha Sood outlines the basics of how hormones are affected by social contact: 'Hormones such as oxytocin, prolactin, vasopressin, estrogen and testosterone influence or are influenced by social interactions. Oxytocin, also known as the "bonding" hormone, is released in some contexts and this can down-regulate cortisol (a stress hormone) release. Oxytocin and dopamine also go hand in hand. The relationship between hormone levels which fluctuate during social interactions changes depending on the nature of the interaction and there is a very complex interplay.'

So our hormonal selves need other people. But it is perhaps no surprise that these cravings for friendships and networks can also fluctuate at different points of the month, subject to both our hormonal ovarian cycle and life stage. Whether we crave a group of friends who want to party, or just want to spend an evening one on one with our nearest and dearest, our social cravings can depend on our hormones as well.

SOCIAL ENGAGEMENT AND OUR MONTHLY HORMONES

For women, commonly the first two phases of your ovarian cycle as estrogen rises are the times you feel most social and more engaged with conversation, as estrogen is a mood-boosting hormone. It can also make you feel more driven in work and even give you better focus and memory. It is the best time of the month to go on dates, as you tend to feel more confident meeting new people in this phase. Use this hormonal high to make important decisions, connections and to network for business. I try where possible to do any public-facing talks, lectures or discussions around this phase of my cycle. It means I tend to be clearer, engaging and more confident in my delivery. Obviously, it's impossible to control the timing of all meetings and work engagements, but if I do have important events when my estrogen is low, I simply ensure I prepare more and give myself longer breaks before and afterwards. Optimising your work and social schedules around how your body and mind might be feeling is a powerful asset in unlocking superpower potential and focus.

The second half of your monthly cycle, post-ovulation (as progesterone rises), can be a good time to be cosier. Your social interactions might feel better if they consist less of big groups and more of one-on-one time with your best friends. It is often in this lower-estrogen phase of the monthly cycle that we crave familiar touch, such as hugs. Hugs (when you remove the risk of a virus) are pure hormonal happiness. They are for most of us the comfort blanket and solution to some of our saddest moments. Often, at our lowest point, it is this simple gesture of kindness that carries us through. The physical touch evokes oxytocin, our bonding hormone. I, for one, am a huge advocate of hugs as medicine and missed them greatly when in solo isolation during the pandemic.

Memory can sometimes be foggier at this time, so workwise it's a good opportunity to take more notes, reflect on thoughts and processes. Avoiding confrontation is key for me in this phase, as I often begin to feel more irritable and have a 'shorter fuse'. For women who suffer from mood swings or more extreme symptoms related to PMDD and endometriosis, it can be a time to focus on calming and more emotional support, ensuring you have both the family and professional backup you need for these more sensitive moments.

Social interaction is as much a practice and cycle as anything else within our lives and, when we understand how our bodies work, we can make the most-informed and best choices. Hormones should not dictate your choices, but understanding them gives you an extra sense and instinct for decision-making in your social life and work. When you tune in to your patterns they simply help inform your navigation through life. The basic question to ask yourself is: What do my hormones need today?

Being able to communicate better how your moods are changing, and in doing so support yourself for the hormonal phase you're in, is like unlocking a new language. It helps friends and family understand that your social choices are there to support your internal health. Friend and family support is also a great way of accessing happy hormones for free, as long as you are careful about spending time with people who make you feel good, not those who make you feel less than or uncertain.

It helps avoid high stress and anxiety at certain times of the month and means that you use your more elevated phases to optimise your work–life balance.

It is important to navigate your relationships for the good of your health, as people can be as much a problem as they can be an antidote. It should be clear by now that stress hormones such as cortisol are not just released when we're under threat of attack. They can be released when we are dealing with the kinds of challenges that other people can bring to our lives. Disagreements and misunderstandings are probably an inevitable reality of being social. While we can't live without our closest friends and family, it often feels like we can't live with them either. However, using the understanding of our hormones to help inform us about when to avoid or address these challenges can be a brilliant tool in negotiation and a powerful way to navigate differences in opinion.

If you experience stress hormones triggered by specific people or social situations, it is worth checking when in your month these triggers are most prominent. See if there is a correlation between when your hormones rise and fall each month and when these people or situations make you feel good or bad. If of course the situation occurs frequently and throughout the month, it is important to ensure you build boundaries between you and the social scenarios that cause sadness over joy.

Ultimately, if people make you feel good, they will help you be healthier. Those who don't will add no long-term value to your physical or mental health, no matter how cool you thought they were when you first met them. Relationships change and so do our bodies: listening to the signals and signs of how our body responds not just to the things we ingest, but also the people we surround ourselves with, is all part of learning our internal and external body's language. That is not to say we should be driven by biological impulse in emotional and social decisions, but it is a factor we can harness in order to better prepare for inevitable changes in our social and life networks.

Although these human relationships, both at home and at work, can sometimes be complex and changing, it helps to keep journals. Keeping a daily or weekly record of how people make you feel, think or act helps frame your view and identify the best ways of overcoming inevitable challenges with your nearest and dearest. It is often from patterns of negative interactions that we find new ways of learning about ourselves and others in social situations. Logging details about relationships can also be a powerful tool for work negotiations. When you keep track of good and bad emotional interactions with your boss or colleagues, you begin to learn how to better navigate certain situations. It helps you to learn best how to manage and approach harder topics or conversations such as when or how to ask for a pay rise. Journals are also an amazing way to look back at your emotional and hormonal growth through life. Having a catalogue of how you feel each day or week, and how other people and places made you feel, is a self-made training manual for how to access DOSE.

FRIENDS FOR LIFE AND FOR CYCLES

Life stages have a huge impact on our emotional and social cycles. Puberty is the time when we are most hormonally volatile, and this can often be reflected in our social relationships. As I outlined in the last chapter, during puberty we are both cognitively and hormonally building pathways of experiences that go some way towards formulating the person we become later in life. Friendship boundaries are often pushed; we see this reflected in social studies of teenage girls and their strong but sometimes volatile bonds.

For women, estrogen fluctuations are linked to emotional and physical changes throughout life; but historically women have been shown to build bonds, groups and relationships in much stronger and longer-term patterns than men.

The shared bonds in women drive DOSE and provide important antidotes to the sometimes extreme emotional symptoms from menopause. Women sharing experiences with close friends and female relatives could be seen as a huge advantage in a world that is being crippled by stress, anxiety and depression.

Social contact and strong friendship bonds have been proven to:

1. Lower blood pressure

2. Reduce BMI and stress

3. Reduce risk of depression

4. Increase lifespan by up to 22 per cent.

In other words, the bonds of friendship are basically the open secret to a long, happy and healthy life.

What women post-menopause also report is their friendships, bonds and clarity on relationships can become much more profound, as reaching a new hormonal life stage opens new doors to new opportunities and new horizons. This can give women a whole new lust for work, careers and relationships. Higher progesterone can also mean women feel less motivated to interact in big social groups and enjoy more focused and meaningful connections.

IRL TO URL SOCIAL NETWORKS

Friendships and close bonds are good for you both in terms of the short-terms effects of social DOSE and in the longer-term effects of building a long and healthy life. Social networks are therefore fundamental to both our own and society's hormonal health. However, social networks have adopted a new persona in the twenty-first century. Since Facebook, the biggest conglomerate social network, launched in 2004, the world of digital social networks has exploded to include Twitter, Instagram, TikTok and Snapchat.

These digital networks have transformed our social lives. They have built a digital way for every human to connect immediately. Although these new social networks have done a lot of good for our ability to connect with those we love and find discourse with like-minded people, they have also fuelled a change in the way that our social hormones and neurotransmitters behave.

The fundamental principles of how these platforms grew was through their ability to build engagement and digital behaviours connected to human desire for affirmation from others. Affirmation obtained through someone attributing value to your posts and comments through digital likes underpins this addictive connection to social networks. Fundamentally we all want to be liked and loved by others. Being liked is a chemical response in our brains and releases two key chemicals of DOSE: serotonin and dopamine. When we meet someone new, either romantically or as a friend, we get a hit; and the same applies when we get affirmation from family, friends or colleagues for good work. This 'gold star' effect – in other words the feeling of being told 'Well done – here is a gold star for your work' – triggers a positive chemical chain reaction inside your brain. But this reaction is not meant to happen every five seconds, as it does if we are getting likes on an Instagram post, and there is an addictive pattern that can form from this kind of frequent affirmation.

This is also paired with the blue light that is emitted from your phone, which as we have already seen has a hormonal impact – increasing cortisol and prohibiting melatonin to stop you sleeping. My own realisation of the power of this addictive affirmation came in my late twenties when I became single for the first time in my adult life. I had been in long-term relationships since I was sixteen and when I was sixteen it was 2003. No digital social networks, no dating apps and no addictive scrolling. On becoming single in 2016, I went from spending most of my free time with either my friends or my partner to having all this extra free time. At first I filled it with reading every book I had missed out on; and then when I began to feel the pangs of loneliness I reached for dating apps and social networks like Instagram to search not just for a partner, but for affirmation. I have to confess I didn't use dating apps for dating. I know this is a terrible thing to admit, as many people I was matching with were probably genuinely looking for a date, but I was finding myself scrolling and matching for the hit of emotional confirmation or ego boost. I was hunting for the DOSE.

During 2017, when I began the research for Moody, I turned my dating life into a mini hormonal science experiment. This was mainly to build on the existing hypothesis that at different stages of my ovarian cycle I would reach for this digital source of DOSE more than others. It was unbelievable how accurately my loneliness and need for affirmation mapped on to my cycle. The craving became almost insatiable in the fourth phase of my month, as my progesterone and vulnerability was higher: I would find myself endlessly scrolling through dating apps, plunged in a deep Instagram hole and posting images to get likes.

In 2018, when we were building the app, we asked hundreds of women if they reached for digital media or apps more at particular times of the month. The results were clear: although there was a use of social media and dating apps throughout the month, there was a direct relationship between the fourth luteal phase and a higher engagement with anything that drove affirmation. In other words, I wasn't the only one who found myself hunting for serotonin and dopamine at the time when women are naturally lower in this final phase of the ovarian cycle.

Breaking addictive or reassurance patterns from apps that give such high and immediate hits of serotonin and dopamine is hard. Although, in theory, being addicted to apps at certain stages of your month seems fairly low-level compared to alcohol, drug or sex addiction, it is still a chemical pathway pattern. And this pattern shares a key component with the neural pathways created from more destructive addictions.

The problem with stimulants, whether chemical or digital, is that they are hard to break out of, as these highs are not like the flow states of consistent and sustained happiness that we get with DOSE: they are spikes. Much like when we eat cake and get both a sugar hit raising our insulin and a dopamine hit, digital addiction is not helping our happiness and fuelling our social interaction through flow; it is providing a hit from which we quickly afterwards feel hungry for more, and that is where the addictive cycle begins and perpetuates.

Technology has changed our brains and hormones. This change is non-reversible, as the technologies that connect our world are here to stay. The impacts of technology on our bodies and brains include shorter attention spans, focus and memory. We now crave the dopamine and serotonin hit from screen light, blue ticks, likes and online affirmations. These biological impacts have changed our brains at an unprecedented evolutionary speed and it is still not known what long-term effects this might have on our brains' and bodies' development. The accelerated pace at which technology began to change behaviours was in January 2007 when the first iPhone landed in our pockets, taking human connectivity and access to information into a new era of immediacy. I am certainly not anti-technology, as I believe it has an important and powerful role to play in how humans can live harmoniously with each other and even with nature. It can help make knowledge, power and even money more democratic. And of course through the Moody app I believe technology can make the science of our hormonal bodies and minds more accessible. These technologies are now moving from infancy to adolescence and therefore it is even more important we evaluate not just how to build them, but also the ethics surrounding them. The addictive behaviours associated with the infinite scrolling of Facebook, Twitter and news feeds wasn't (I assume) the intention of innovators like Jack Dorsey and Mark

Zuckerberg, but this is one of the many human impacts. As with any industry, it evolves; and what the next generation of technology is responsible for is learning from past mistakes and not repeating bad behaviours. Innovating in how to build technologies that provide healthy relationships and still connect people with the world and those around them is a technological future I am proud to be part of: an industry that is looking ahead at how to help people feel more well, get more educated and be happier. No simulated or digital world will ever be able to replace the power of human touch, so spending time looking at this new era of more emotionally and biologically responsible technologies is important for anyone in the sector.

When building the Moody app, we researched addictive behaviours and chemical interactions to ensure we were creating technology that focused on connecting people with themselves, not perpetuating addictive behaviour linked to external affirmation from others. Health habits can be built without addiction. You have the power inside you to give yourself affirmation and happy hormones, and this is what we wanted to build: a new generation of technology that focused on helping people re-educate and unlock their health and happiness. There is a way of finding benefits from social and digital technologies so long as we understand how they affect our brains and bodies.

The principles we followed for Moody can also be a good basis for you to build healthy relationships with all the apps and technology you use:

1. **No infinite scroll**. Limiting the number of news or social platforms you have available on your phone can help stop the compulsion not just to scroll, but to jump from one scroll to another.

2. **Focused time frames for use**. Having specific times in the day where you limit or use apps can avoid the addictive triggers, but it is also powerful to have a switch-off from

all screens by 9 p.m., as this will curtail the effect of blue light on your sleep routine. This simple act can transform your mental and physical health.

3. **No likes**. Likes and digital affirmation can become addictive. Understanding this means you can check in with yourself if you ever find yourself reaching for apps and likes when you feel low. Try and search out apps and platforms that don't rely on likes and ticks.

Moderating the amount of time you spend scrolling online, reading media and communicating via technology is one solution to giving your brain and body rest and recovery. It's called a digital detox, but there is a truth to this buzz word. Building a balanced and less addictive routine with technologies is an important part of not just reducing anxiety and therefore cortisol, but also reducing the amount of blue-light exposure you have, which we have seen has a hormonal impact.

It is for this reason that Moody is designed around a daily check-in, and the idea of having a moment in your day when you think about your hormones, moods and symptoms, as a guide for how to navigate the IRL world around you. Our social lives should not be determined by digital experiences or even by our hormones. Instead we should be using and choosing these assets as tools to help navigate, educate and inspire us to be the best version of ourselves for ourselves. As when you are in a flow state of happiness, this transfer is shared when you then engage with friends and family. Happiness is infectious.

HORMONES FOR IMMUNITY

I wrote this book during a global pandemic. Covid-19 has been the most serious pandemic since the Spanish flu in the early twentieth century. What changed dramatically between 1920 and 2020 was technology, but what hasn't changed is the human vulnerability to a new fast-spreading virus. The world has become richer and more connected, but nothing prepared the economy and civilisation for a natural and invisible pathogen.

Experts predict that pandemics will become more frequent due to industrial farming, which is breeding animals in environments that become hotbeds for aggressive viral strains. Human populations are growing and the rich nations continue to consume food and resources at vast rates. This is having a direct impact on the natural cycles of our world's ecosystems and, in turn, on our bodies. This is not just about the risk from a deadly virus, but also the risks of air pollution and consumption of toxins from pesticides in the food we eat or microplastics from water we drink. The sad truth of world cycles of consumption, and the impact they are having on human health, is that the poorest and most vulnerable are most affected, as the billions of people living under the poverty line across the world are being exposed to much more extreme effects from drought, famine and lack of sanitation.

If we start to think about the Earth as a living being, we can begin to see why as a human population we must support each other. Humans are only one part of the planet's ecosystem. To support and thrive on Earth we must live alongside all the other inhabitants. The world we call home has its own cycles and patterns. Some we understand and some we are yet to discover, but what we know is that there is a balance and when that balance is disrupted it has a knock-on effect on all existing ecosystems. This equilibrium is clear when you realise the delicate but integral function of biodiversity for the planet's survival.

When you put too much toxic stress, such as CO_2, into the system, there are signals and signs that the Earth becomes more volatile and reactive. Just as when our bodies are flushed with cortisol and we see aggressive symptoms such as fatigue, weight fluctuations or skin flare-ups, the Earth has its own signals in the form of weather extremes such as cyclones, flooding and droughts. Tuning in to these changes in the world's cycle can help us look at both the individual and collective impacts these chemical stressors are having.

Every person on this planet has the power to change cycles, both individually and by joining forces collectively. It starts with education. We can't make reverse changes to our environment if we don't know the systems behind why and how it became so toxic.

When the coronavirus hit, immunity became a top trending word in Google search; and Moody as a technology company went from a fringe hormone app to a global solution, as hormones are fundamental to immunity responses. The key change that occurred was a realisation by the world's populations that they had the future of their health in their own hands – not just because the most powerful defence against the virus was washing your hands, but also because every human became aware of how their social interactions could be a risk of infection. The news was filled with how to stay safe and the consistent theme was stay away from groups and mixing with others, but the issue with this enforced isolation is, although it prevents the short-term spread of a deadly virus, it doesn't keep our mental health in check: social interactions drive happy hormones and help us build immunity, the very defence system we need for protection from the virus. Therefore, society began to trial new systems of how to protect people, while also ensuring our bodies didn't become victim to more profound stress and anxiety disorders driven by isolation.

Immunity is supported by hormones both in the production of immune cells, but also in keeping our overall endocrine systems functioning effectively, keeping our white blood cells healthy and moving. The right amount of happy hormones means stronger and more resilient immune function and cell cleansing and formation. As we have seen, the key gland involved in the immune system is the thymus. This is positioned in the chest and produces several hormones that contribute to immune-cell development:

Thymopoietin is a key part of the process whereby immune cells differentiate into different types.

Thymosin elevates the immune response as well as stimulating pituitary hormones such as growth hormone, which in turn contribute to our growth and life stages.

Thymic humoral factor acts similarly to thymosin, but increases the immune response to viruses in particular.

These three hormones are part of a stimulation process that creates and maintains our immune cells, which are needed to attack and defend against a virus such as Covid-19. Immunity cells are split into two types: T lymphocytes and B lymphocytes (in shorthand T and B cells):

B cells. These are your defence system, as they produce antibodies that are directed at specific invaders.

T cells. These are the active attack cells, fighting off bacteria, viruses, cancer cells and any other foreign body.

The immune system is not just a case of the thymus working in isolation to create strong immunity responses: each gland works in harmony, to regulate and maintain the natural production of these immunity-stimulating hormones. This is why anyone with underlying health conditions is more vulnerable to a virus like Covid-19, as their body is already fighting and using immunity cells to defend against existing illness.

Dr Minisha Sood describes again what we have already seen throughout this book: the way in which all our hormonal systems are interconnected, and why this means that stress impacts on our levels of immunity: 'There is a close connection between circadian rhythms (sleep–wake cycles and other cues) and the stress response. Stress hormones include cortisol and adrenaline. Two-way communication between the immune system and the hypothalamic–pituitary–adrenal (HPA) axis exists. Certain chemicals, called cytokines, released by immune cells can directly stimulate the production of stress hormones. Cortisol, a glucocorticoid hormone, can acutely change the immune response, shifting it from a pro-inflammatory response to one that is anti-inflammatory. Chronic stress, on the other hand, is thought to somehow suppress the immune system though this mechanism is not well understood. In addition, disruptions in circadian rhythm can also introduce stress and negatively impact immune function.' Although the science and language may sound complicated, the principle is simple: stress has a direct effect on the body's ability to defend itself against illness. This means the more stressed you are, the more at risk you are of falling ill.

As with all our cycles, no one person's system is identical: we each have a unique chemical combination inside us, making some of us more susceptible to illness than others. However, there are things we can do to support this important system. There are even some superpower supplements and natural products that help our human systems stay strong. Having a balanced and healthy hormonal diet and exercise routine connected to your monthly cycle is one part of supporting your body's ability to build immune cells. The following foods and supplements can also help:

Zinc – found in lean meat, seafood, milk, wholegrains, beans and nuts.

Iron – found in lentils, spinach, tofu and white beans.

Vitamin C – found in broccoli, kale, citrus fruits, strawberries and tomatoes.

Vitamin E – found in nuts, seeds, green leafy vegetables, avocados and shrimps.

Vitamin D – found in sunshine (the key way of obtaining it). It is also present to an extent in egg yolks, salmon and mushrooms.

Echinacea – a flowering plant that has proven benefits in assisting immunity and increasing the number of white blood cells in your system and supporting antibody formation. It is a supplement well worth taking every day, as you never know when you might come in contact with a virus or germs that could cause you to fall ill.

There is also some important work being done on how medicinal mushrooms can provide supplementation to our body's ability to build and maintain high levels of B and T cells. Mushrooms are proteins and contain some vitamin D. Specific strains can aid your body in building antibodies and 'attack cell' development for fighting off an illness or virus.

Research by Johns Hopkins University and Paul Stamets has shown how mushrooms can support immunity and healing from chronic illness such as cancer. The study demonstrates that it is about taking these products *alongside* your prescribed drugs, not one or the other. This is an example of where natural solutions and pharmaceuticals can sit harmoniously together: to opt for one without the other is setting yourself up for risks. And I would never advise choosing any healing route without speaking to a doctor first. Certain mushrooms may positively influence the gut microbiota, improving protection against pathogens and supporting natural metabolic rates. However, natural solutions and mushrooms can come in different strains, with the quality and standard of the mushroom contributing to how effective it is. Much like any fruit or vegetable, they have more nutritional value and taste better when they have been grown in the right soil, with the right lights and temperature levels and arguably with a little bit of love. Mass-farming anything can strip these natural-born healers of their most potent properties, so being aware of by whom and how they have been produced will give you an indication of whether they will provide the best-possible boost for your health and immunity.

The top mushrooms for immunity are:

Turkey tail (*Coriolus versicolor*). This mushroom gets its name from the feather-like pattern on its surface. It contains two key antioxidants: phenol and flavonoid. Both promote immune system health by reducing inflammation and stimulating the release of protective compounds. Turkey tail has been used therapeutically for treating cancer and acquired immunodeficiency syndrome (AIDS). Contemporary research shows the anti-tumour and anti-metastasis effects have in patients reduced the risk of developing secondary cancerous growths.

Reishi (*Ganoderma lingzhi*). Reishi is commonly referred to as the mushroom of 'immortality' and 'happiness'. It has been shown to support sleep, reduce stress and lessen fatigue, all symptoms linked to various hormone imbalances. The health effects of this mushroom may be a result of its ability to regulate microbiota composition, as the polysaccharides found in reishi demonstrate prebiotic effects and may increase counts of beneficial bacteria in the gut. Its properties have been shown to directly affect cellular health and boost immunity.

Cordyceps (*Ophiocordyceps sinensis*). Cordyceps is a caterpillar fungus, so not technically a mushroom, but still within this healing family. It is associated with increased activity of cytotoxic T cells, AKA natural killer immune cells.

Maitake (*Grifola frondosa*). Maitake is both a culinary and medicinal mushroom that has shown anti-cancer activity in animal studies. It is the component proteoglycan found within maitake that has been associated with its immune-stimulating effects. Polysaccharides also found in maitake have been shown to exert anti-viral activity against hepatitis B and human immunodeficiency virus (HIV).

Lion's mane (*Hericium erinaceus*). The name comes from this mushroom's white fur-like texture. Studies have demonstrated that lion's mane reduces mild anxiety and depression symptoms and even repairs nerve and neurological tissue. Its cognitive health benefits have also shown results in studies on Alzheimer's and Parkinson's disease. It is specifically linked to gut support and intestinal immunity, which is crucial in preventative health.

Shiitake (*Lentinula edodes*). Shiitake is one of the most popular mushrooms worldwide, not just for its medicinal properties, but it is also delicious to eat. It has historically been used to treat conditions such as the common cold. The species contains polysaccharides, terpenoids, sterols and lipids, all of which contribute to immunity boosting.

What is key to immunity is that you can't guard against every eventuality, but it is a natural defence system which can be supported through the decisions you make in food and lifestyle. As humans we are vulnerable to all sorts of factors, but by having the basic knowledge of how these cycles and systems work, you can give yourself a head start. Immunity, much like menstruation, sleep and metabolism, is unique to you and therefore learning how and what works for you is about listening and tuning in to the routines, rituals and foods that make your body feel strong, give you energy and help you recover from a common cold quickly. This will then give you the best foundations for a body that is at least prepared for what life will inevitably throw at you, in all its unpredictability.

THE WIDER PICTURE – GLOBAL SYSTEMS

There is a relationship between our bodily cycles and the wider social and political world. Understanding the impact of our way of life on the Earth's climate and resources is not just to draw the links between nature and humanity, but also to show there is a more important literal connection between these global changes and our bodies. Our bodies are being hormonally impacted by these macro cycles of climate changes through pollution and political stress, whether we like it or not. The impact of living in a more polluted world is affecting our hormones, as these toxins are absorbed and need to be excreted from the body's system. This means they affect how well all the hormone cycles of the body work – from metabolism through sleep to stress. They are having a daily impact on our immunity and therefore on our body's ability to fight off illness. According to neuroscientist and researcher Araceli Camargo, who has worked with Moody on the impact of environments on our hormones: 'A person can only have agency over their health when they are part of a supportive ecosystem, which includes the places they inhabit. What this means is that a person's hormonal health is tied to the health of the environment they live in. This includes the agency they have to achieve hormonal health. If a person lives in an area with high levels of air pollution, this will impact the stress response, which in time will affect all aspects of a person's biological systems, including endocrine.'

This can be best understood by unpacking some of the social systems involved. It might seem mad, but there is a direct link between money, food, politics and our hormonal bodies. These systems are not natural world orders to keep our planet and bodies healthy; they are designed by humans to fuel economic growth. There is a basic equation that can be drawn. Healthy and happy humans + healthy planet = long-term survival of both planet and human populations. Seems fairly simple, but humans have built very complicated economic and political cycles that mean this equation becomes harder and harder to realise. And, as Araceli points out, this does not affect all women equally, due to social and economic factors: 'Black and indigenous women disproportionately live in biological in-equality, which is defined as a person's disproportionate chronic exposure to environmental (air pollution, inadequate housing) and psychosocial stressors associated with poverty (income, housing insecurity). The experience of this phenomenon can in turn put them at higher susceptibility for disease and biological dysregulation (hormonal/metabolic/digestive). It is important to note that susceptibility is not the same as genetic determinism as espoused by some ill-informed scientists.'

One thing the 2020 pandemic highlighted is the vulnerability in our global health systems. Our current systems have been designed to treat symptomatic illness, which means you only go to the doctor when you have a problem. What has informed this health system is a business model. Rather than educating us all on how to stay healthy, it treats symptoms. The business model of pharmaceutical companies is one of supplying drugs and treating problems when they happen. Keeping people well would mean less money for an industry driven by symptomatic treatments and patented drugs. It's a cycle of money that feeds the cycle of illness and is highly illogical for creating human happiness and health.

There has been a historical global divide in health education, based on two camps: a more traditional, 'natural' philosophy which focuses on prevention; and a more 'modern', pharma-ceutical-based approach which focuses on cures. The reality is we need both prevention and cures to support human health. A positive shift in the last ten years is that this divided approach is beginning to meet in the middle. The future is not about choosing which side you sit on, but about bridging prevention with effective innovation in medical treatments. Keeping people's bodies healthy on a daily basis to prevent illness occurring and reducing the impact of illness through pharmaceutical cures, where necessary: this seems like the perfect united front.

A model that helps keep us healthy is one that keeps us more connected to our own bodies. It is one where we possess the innate knowledge about when to ask for medical help. Tuning in to your daily health and well-being makes you an expert on how your body feels and therefore more effective in knowing how to utilise the luxury of healthcare systems and medical support if you have access to them.

When the pandemic hit, the health and wellness industry switched overnight in March 2020 from a vanity sector to a cornerstone of human survival. This is because the wellness industry is hugely beneficial to one key system – our immune responses – and it is within our own power to support this.

CONSUMING THE WORLD

It should be clear by now that what we ingest into our bodies – in the form of food, or pollutants, or even stress – will have an impact not only on our own health and hormones, but on the systems and networks that make up our world. And nowhere is this clearer than in the realm of junk food. Low-cost food manufacturing is another example of a Western cultural and economic system that has fuelled this misalignment between our own bodies and the world. We don't think of food as medicine for our body, to help balance hormones or support immunity. Often, we eat what we eat for emotional reasons. This can be positive, making us happy and building social connections over a delicious meal. But it can have less beneficial consequences as well.

All the food we eat has an impact on our health, both good and bad. And we have also built food rewards that cause more harm than they solve. Take the industry that surrounds mass-produced junk foods. I always find it fascinating that no one seems to have noticed that we have an entire industry of food 'junk' and although we know it is rubbish or trash, we still ingest it. The reason we ingest it, is not a lack of self-control, but that this industry of food is chemically manufactured to be highly addictive. It's just governments seem to class these addictive chemical drugs as legitimate drugs and allow them to be fed to children, creating the lifelong chemical addiction and dependents we see across society. The foodstuffs involved create illnesses such as obesity and diabetes. They inhibit our natural metabolism, through high saturated fats and refined sugars. And they can be addictive. And that is just the impact on us at a personal level. The industry of 'junk food' is a global economic system and superpower. Low-cost food uses mass-farmed meats, vegetables and grains, as well as plastic packaging, meaning it is also having a broader impact on air pollution and plastics in water.

Melissa Hemsley put it well when describing what she aims for in the food she consumes: 'For me, I also like to ensure I don't just make meals that just look good on Instagram, but also where I feel good about where I bought or sourced the ingredients. Eating food that hasn't had a negative environmental impact is important for me to enjoy the meal I have made. Feel-good is also about being conscious of our food choices, both for ourselves and for the planet, so everyone can continue to live and eat happily together.'

I am not exempt from buying 'junk food' – I have loved it at times – but my previous decision to indulge in this system was based merely on weighing up the excess fats and sugars in my own body. I had never considered the other, bigger systems having a negative impact on the world or my life until someone explained to me how the simple act of buying junk food has a much more profound effect than just me eating and being unhealthy. I had never considered the carbon impact of eating cheap junk food and therefore the impact this toxicity was having on my long-term health and happiness.

Even if the price was low in cost to me in the immediate, the costs that aren't labelled suddenly started to become part of my decision-making processes. Why has no one added into the price of junk food the true cost of pollution or illness? Obesity alone in America costs $190 billion each year, which is 21 per cent of the nation's annual medical spend. Imagine if this bill was then sent back towards the food-manufacturing compan-ies. The problem with this system of mass-produced food is that politics and economics are not fair in how they attribute value or cost. The system was designed to favour cheap products with low nutritional value, driving up scales of waste and CO_2 and driving down the overall health and happiness of citizens. And it rewards 1 per cent of wealthy citizens and leaves 99 per cent of the population without access to basic food and water supplies.

There have been some incredible documentary films, such as *Fed Up* by Stephanie Soechtig, which have addressed these topics of food consumption, politics and corporations to effect change. Campaigners continue to try to change this cycle, but it's a complicated political system not set up to respond quickly.

What has been fundamentally missed from our education and social systems is an appreciation for the value of health, as something that we all have the power to harness and tune in to. Not just for those who have access to healthcare or wellness routines, but through the education of how food, lifestyle and the world around us all have an impact on our body's ability to operate effectively. Ultimately, what we put into our system, both actively through food and passively through our environ-ment, makes a difference to our health and happiness. This information should be accessible to everyone and is arguably more important than an education focused on world wars and curriculums designed by governments who are often divided by economics, history and politics.

This leads on to the final key connection between world cycles and our hormonal health, and that is how our brains and bodies learn hormonal responses from past and present events both in our lives and from the social worlds and spaces around us. Every action or experience in life becomes information that is received through our brains. Our mind and body connection is hormonal, as all information passed from brain to body is communicated via a series of neurological and hormonal reactions.

Every experience, every interaction, every life goal and every life stage has a process of hormonal imprints. Mental health is often dismissed by ourselves and by the world, as it has no tangible wound or physical symptom to identify it. We often leave the mental signs unattended for so long that they manifest as physical signs and, by this stage, the stress, anxiety and depression may have taken a much deeper hold. It is not just important to support hormonal health as a way of accessing hormones that can help you to live a healthy physical life, but also to promote more long-term mental health and access to happiness and joy.

We must employ agency, not just in protecting ourselves, but also in being kind and mindful of how we consume, present and talk to each other through media and in real life. Reducing the potential impacts of chronic stress and anxiety is critical to hormonal health.

YOU HAVE THE POWER TO CHANGE THE WORLD

When I began my journey of hormonal re-education, what struck me was how our hormones don't just affect our mind and bodies, but are connected to all the other systems and cycles around us – from relationship stress, to money worries and even pollution levels. They are a system that can be tapped into and help us all become more literate in our own health and the health of others. This then means we can become more aware of the impacts our choices have on the world around us, both locally and globally. Collective human happiness can sometimes sound like a utopian vision for a world that humans have often spent more time being careless of than protecting. What is powerful to remember is the importance and value everybody has in changing these cycles.

The goal of changing the world can feel abstract, exhausting or just too much to take on, and far removed from problems we seem to face on a daily basis. However, it can be achieved, and everyone can contribute by starting with their own health and well-being. The negative impacts of human social systems disproportionally affect those in less-developed economic nations and those with lower income. The impact of pollution, however, affects *all* of us and *all* of our bodies. If you have the privilege of capital, spending it wisely on your own health and on products or services that have a positive impact on the world's ecosystems will go a long way to counteracting some of these cycles of poverty, inequality and pollution. We are all affected by a world that has become so fast-moving and fast-consuming that we ingest hormonally-disrupting chemicals through the food, water and air we breathe.

The planet lives and breathes just like all the organisms that inhabit it. World cycles and systems are not permanent: they have been built over the last fifty years and can be broken down and adapted. They are human-made; they can be human-changed. I am not trying to convert everyone into being a purely altruistic citizen. It would be unrealistic to imagine that we should all begin to exist outside our own ego and in the pursuit of human happiness. Human beings are by design selfish, as we are designed to survive.

Survival, however, is about others and the health of the planet. Happier communities, civilisations and environments means your body will be working at its optimal level. Cycles are not just about periods, metabolism, sleep or sex; they are about how everything living has a beginning, middle and end. Your hormones are a power cycle. When you tap into it, understand it and even elevate it, you can begin to relate to other women, people and places. This knowledge can give you a sixth sense. Small acts of kindness, confidence and support can create a happy hormonal domino effect for everyone around you. Every great superhero story starts with how that hero acquired their powers. And now you know: you have the powers inside you.

TAKEAWAYS

- We need human contact and the hormones released when we see friends, family and networks to stay healthy.

- Social impulses can change for some women across their menstrual cycle, as fluctuations in estrogen and progesterone can make you want to socialise more or less.

- Digital and social media can create addictive behaviours and patterns, evoking more stress hormones than happy hormones. Physical contact is key for friendships and social time.

- Stress hormones inhibit and can cause a negative impact on our immune systems.

- Understanding your hormones and all the cycles involved in your body can connect you better with yourself and help you connect better with the world and people around you. It can give you superpowers for life, love, work and more.

READING LIST

Below is a non-exhaustive list of books to further your learning:

Bessel van der Kolk, *The Body Keeps the Score*

Melissa Hemsley, *Eat Happy*

Mihaly Csikszentmihalyi, *Flow*

Giulia Enders, *Gut: The Inside Story of Our Body's Most Underrated Organ*

Dr Anita Mitra, *The Gynae Geek*

Kimberly Wilson, *How to Build a Healthy Brain*

Dr Angela Saini, *Inferior*

Caroline Criado-Perez, *Invisible Women*

Charlie Howard, *Misfit*

Dr Philippa Kaye, *The M Word*

The Hotbed Collective, *More Orgasms Please*

Maisie Hill, *Period Power*

Jill Shaw Ruddock, *The Second Half of Your Life*

Dr Tara Swart, *The Source*

Dr Gen Gunter, *The Vagina Bible*

NOTES

Abdallah, C. G. et al. (2013), 'A pilot study of hippocampal volume and N-acetylaspartate (NAA) as response bio-markers in riluzole-treated patients with GAD', *European Neuropsychopharmacology* [online], 23 (4), 276–84.

Amianto, F. et al. (2013), 'Brain volumetric abnormalities in patients with anorexia and bulimia nervosa: a Voxel-based morphometry study', *Psychiatry Research: Neuroimaging* [online], 213 (3), 210–16.

Andrews, C. R. (2014), 'Punk has a problem with women. Why?', *Guardian* [online], 3 July. Available from: http://www.theguardian.com/music/musicblog/2014/jul/03/punk-has-a-problem-with-women-why (accessed 1 August 2018).

Anon (n.d.), 'A broken idea of sex is flourishing. Blame capitalism', Rebecca Solnit, *Guardian* [online], 12 May. Available from: https://www.theguardian.com/commentisfree/2018/may/12/sex-capitalism-incel-movement-misogyny-feminism (accessed 27 June 2018).

Anon (n.d.), 'A systematic review of the effects of hormone therapy on psychological functioning and quality of life in trans-gender individuals', *Transgender Health* [online]. Available from: https://www.liebertpub.com/doi/full/10.1089/trgh.2015.0008 (accessed 27 June 2018).

Anon (n.d.), *About | National Organization for Women* [online]. Available from: https://now.org/about/ (accessed 3 August 2018).

Anon (n.d.), *About | Everyday Sexism Project* [online]. Available from: http://everydaysexism.com/about (accessed 1 August 2018).

Anon (n.d.), *About the FED*, Feminist economics department [online]. Available from: http://feministeconomicsdepartment.com/the-fed/ (accessed 1 August 2018).

Anon (2017), 'BBC 100 Women: who is on the list?', BBC [online], 1 November. Available from: https://www.bbc.co.uk/news/world-41380265 (accessed 27 June 2018).

Anon (2015), *Black, Asian and Minority Ethnic (BAME) Communities* [online]. Available from: https://www.mentalhealth.org.uk/a-to-z/b/black-asian-and-minority-ethnic-bame-communities (accessed 28 July 2018).

Anon (n.d.), 'Constellating queer spaces', *Urban Omnibus* [online]. Available from: https://urbanomnibus.net/2018/02/constellating-queer-spaces/ (accessed 27 June 2018).

Anon (n.d.), *Encyclopedia of Human Nutrition – 3rd Edition* [online]. Available from: https://www.elsevier.com/books/encyclopedia-of-human-nutrition/allen/978-0-12-375083-9 (accessed 3 August 2018).

Anon (2008a), *Estrogen and progestogen use in postmenopausal women: July 2008 position statement of the North American Menopause Society* [online], 15 (4 Pt 1), 584–602.

Anon (2008b), *Estrogen and progestogen use in postmenopausal women: July 2008 position statement of the North American Menopause Society* [online], 15 (4 Pt 1), 584–602.

Anon (n.d.), *FT-New-Femininity-RP.pdf* [online]. Available from: https://jwt.co.uk/uploads/pdfs/FT-New-Femininity-RP.pdf (accessed 3 August 2018).

Anon (2010), *Gender Equality Results Case Studies: Bangladesh*.

Anon (2018), 'Gender identity needs to be based on objective evidence rather than feelings', *The Economist* [online]. Available from: https://www.economist.com/open-future/2018/07/03/gender-identity-needs-to-be-based-on-objective-evidence-rather-than-feelings (accessed 2 August 2018).

Anon (n.d.), *GLOBOCAN Cancer Fact Sheets: Cervical Cancer* [online]. Available from: http://globocan.iarc.fr/old/FactSheets/cancers/cervix-new.asp (accessed 25 July 2018).

Anon (2017), 'How beauty giant Dove went from empowering to patronising', *Guardian* [online], 15 May. Available from: https://www.theguardian.com/fashion/2017/may/15/beauty-giant-dove-body-shaped-bottles-repair-damage (accessed 3 August 2018).

Anon (n.d.), *How can I tell when I'm ovulating?* NHS [online]. Available from: https://www.nhs.uk/common-health-questions/womens-health/how-can-i-tell-when-i-am-ovulating/ (accessed 22 July 2018).

Anon (n.d.), *How to Design a City for Women* [online]. Available from: https://www.citylab.com/transportation/2013/09/how-design-city-women/6739/ (accessed 16 July 2018).

Anon (n.d.), *Information on Estrogen Hormone Therapy, Transgender Care* [online]. Available from: https://transcare.ucsf.edu/article/information-estrogen-hormone-therapy (accessed 27 July 2018).

Anon (n.d.), *Information on Testosterone Hormone Therapy, Transgender Care* [online]. Available from: https://transcare.ucsf.edu/article/information-testosterone-hormone-therapy (accessed 27 July 2018).

Anon (n.d.), *Introduction to the Review*, Equality and Human Rights Commission [online]. Available from: https://www.equalityhumanrights.com/en/anghydraddoldeb-traws-wedi%E2%80%99i-adolygu/introduction-review (accessed 2 August 2018).

Anon (n.d.), *Jo-Anne Bichard* [online]. Available from: http://arts.brighton.ac.uk/staff/jo-anne-bichard (accessed 16 July 2018).

Anon (n.d.), *Keeping your vagina clean and healthy*, NHS [online]. Available from: https://www.nhs.uk/live-well/sexual-health/

keeping-your-vagina-clean-and-healthy/ (accessed
1 August 2018).

Anon (n.d.), 'Key statistics about women and mental health',
Counselling Directory [online]. Available from: https://www.
counselling-directory.org.uk/women-and-mental-health-stats.
html (accessed 25 July 2018).

Anon (2009), 'Lack of healthcare worsens women's life quality:
WHO' [online], 9 November. Available from: https://www.reuters.
com/article/us-women/lack-of-health-care-worsens-womens-
life-quality-who-idUSTR E5A85BB20091109 (accessed 25 July
2018).

Anon (n.d.), *lgbt-in-britain-trans.pdf* [online]. Available from:
https://www.stonewall.org.uk/sites/default/files/lgbt-in-britain-
trans.pdf (accessed 2 August 2018).

Anon (n.d.), 'Long-term consequences of anorexia nervosa',
Maturitas [online]. Available from: https://www.maturitas.org/
article/S0378-5122(13)00125-4/fulltext (accessed 3 August 2018).

Anon (n.d.), *Making Infrastructure Work for Women and
Men: A Review of World Bank Infrastructure Projects
(1995–2009)* [online]. Available from: http://siteresources.
worldbank.org/EXTSOCIALDEVELOPMENT/Resources/
244362-1265299949041/67 66328-1270752196897/
Gender_Infrastructure2.pdf (accessed 27 June 2018).

Anon (n.d.), 'New cancer case number among U.S. women 2018',
Statistic [online]. Available from: https://www.statista.com/statis-
tics/268502/us-number-of-new-cancer-cases-among-women/
(accessed 25 July 2018).

Anon (2017), 'Nike takes a stance for a new femininity', *Sleek
Magazine* [online]. Available from: http://www.sleek-mag.
com/2017/05/23/nike-cortez/ (accessed 24 July 2018).

Anon (n.d.), *Sexual and reproductive health and rights*, OHCHR
[online]. Available from: https://www.ohchr.org/EN/Issues/

Women/WRGS/Pages/HealthRights.aspx (accessed 27 June 2018).

Anon (n.d.), *Ovarian Cancer*, World Cancer Research Fund UK [online]. Available from: https://www.wcrf-uk.org/uk/preventing-cancer/cancer-types/ovarian-cancer (accessed 25 July 2018).

Anon (n.d.), *Playbour*, Work, Pleasure, Survival [online]. Available from: http://workpleasuresurvival.org/index.php (accessed 1 August 2018).

Anon (n.d.), *Premenstrual syndrome (PMS) – symptoms and causes* [online]. Available from: http://www.mayoclinic.org/diseases-conditions/premenstrual-syndrome/symptoms-causes/syc-20376780 (accessed 22 July 2018).

Anon (n.d.), *Punk Heaven for Little Girls*, BBC [online]. Available from: https://www.bbc.co.uk/programmes/p02sxfp6 (accessed 2 August 2018).

Anon (n.d.), *Reform of the Gender Recognition Act 2004* [online]. Available from: https://www.gov.uk/government/consultations/reform-of-the-gender-recognition-act-2004 (accessed 3 August 2018).

Anon (n.d.), *Sexual Agency of Women* [online]. Available from: http://www.bl.uk/learning/histcitizen/sisterhood/clips/sexuality-love-and-friendship/sexual-pleasure-sexual-rights/143182.html (accessed 3 August 2018).

Anon (n.d.), *Social Mobility* [online]. Available from: https://www.britannica.com/topic/social-mobility (accessed 31 July 2018).

Anon (n.d.), *Social Problems in Pregnancy*, ScienceDirect [online]. Available from: https://www.sciencedirect.com/science/article/pii/S1472029906700650 (accessed 1 August 2018).

Anon (n.d.), *Stress: Concepts, Cognition, Emotion, and Behavior – 1st Edition* [online]. Available from: https://www.elsevier.com/books/stress-concepts-cognition-emotion-and-behavior/fink/978-0-12-800951-2 (accessed 1 August 2018).

Anon (n.d.), 'Structural neuroimaging studies in major depressive disorder: meta-analysis and comparison with bipolar disorder', *Depressive Disorders*, JAMA Network [online]. Available from: https://jamanetwork.com/journals/jamapsychiatry/fullarticle/1107416 (accessed 1 August 2018).

Anon (n.d.), *The 96th Street divide: why there's so much diabetes in Harlem* [online]. Available from: diabetes-in-harlem-new-york-obesity (accessed 28 July 2018).

Anon (n.d.), 'The brain's default mode network', *Annual Review of Neuroscience* [online]. Available from: https://www.annualreviews.org/doi/full/10.1146/annurev-neuro-071013-014030#_i2 (accessed 1 August 2018).

Anon (n.d.), 'The fourth wave of feminism: meet the rebel women', *Guardian* [online]. Available from: https://www.theguardian.com/world/2013/dec/10/fourth-wave-feminism-rebel-women (accessed 3 August 2018a).

Anon (n.d.), 'The fourth wave of feminism: meet the rebel women', *Guardian* [online]. Available from: https://www.theguardian.com/world/2013/dec/10/fourth-wave-feminism-rebel-women (accessed 1 August 2018b).

Anon (n.d.), *The Great British Public Toilet Map* [online]. Available from: https://greatbritishpublictoiletmap.rca.ac.uk/ (accessed 16 July 2018).

Anon (n.d.), 'The subversive power of hyper-feminine fashion', *i-D* [online]. Available from: https://i-d.vice.com/en_us/article/zmx558/the-subversive-power-of-hyper-feminine-fashion (accessed 3 August 2018).

Anon (n.d.), 'Transgender architectonics by Lucas Crawford', *Society & Space* [online]. Available from: http://societyandspace.org/2017/04/25/transgender-architectonics-by-lucas-crawford/ (accessed 27 June 2018).

Anon (n.d.), 'Two sexes are not enough', *NOVA* [online]. Available from: http://www.pbs.org/wgbh/nova/body/fausto-sterling. html (accessed 30 July 2018).

Anon (n.d.), 'Two women who pioneered user-centered design', *ACM Interactions* [online]. Available from: http://interactions. acm.org/archive/view/november-december-2013/two-women-en-who-pioneered-user-centered-design (accessed 16 July 2018).

Anon (n.d.), *ukpga_19670087_en.pdf* [online]. Available from: https://www.legislation.gov.uk/ukpga/1967/87/pdfs/ ukpga_19670087_en.pdf (accessed 3 August 2018).

Anon (n.d.), *Understanding Gender*, Gender Spectrum [online]. Available from: https://www.genderspectrum.org/quick-links/ understanding-gender/ (accessed 22 July 2018).

Anon (n.d.), 'University lecturer criticised after declaring "trans women are still males with male genitalia"', *PinkNews* [online]. Available from: https://www.pinknews.co.uk/2018/07/06/ university-lecturer-says-trans-women-are-still-males-with-male-genitalia/ (accessed 2 August 2018).

Anon (n.d.), *WALK-ON* [online]. Available from: https://issuu. com/stereographic/docs/walkon_for_issuu (accessed 27 June 2018).

Anon (n.d.), *We Are Moods – Women's Health & Hormones*, We Are Moody [online]. Available from: https://wearemoody.com/ (accessed 27 June 2018).

Anon (n.d.), *What are some common complications of pregnancy?* [online]. Available from: http://www.nichd.nih.gov/ health/topics/pregnancy/conditioninfo/complications (accessed 1 August 2018).

Anon (n.d.), *Gender*, WHO [online]. Available from: http://www. who.int/gender-equity-rights/understanding/gender-definition/ en/ (accessed 1 August 2018).

Anon (n.d.), *Gender and women's mental health*, WHO [online]. Available from: http://www.who.int/mental_health/prevention/genderwomen/en/ (accessed 25 July 2018).

Anon (n.d.), *WHO-MSD-MER-2017.2-eng.pdf* [online]. Available from: http://apps.who.int/iris/bitstream/handle/10665/254610/WHO-MSD-MER-2017.2-eng.pdf?sequence=1&i sAllowed=y (accessed 1 August 2018).

Anon (1963), 'Women: a new femininity', *Time* [online]. Available from: http://content.time.com/time/magazine/article/0,9171,829840,00.html (accessed 24 July 2018).

Anon (2018), 'Women's reproductive health is the most neglected thing in our society', *Huffington Post* [online]. Available from: https://www.huffingtonpost.in/india-development-review/women-s-reproductive-health-is-the-most-neglected-thing-in-our-society_a_23385135/ (accessed 28 July 2018).

Anon (n.d.), *Women's Suffrage Movement Archives*, University of Manchester Library [online]. Available from: http://www.library.manchester.ac.uk/search-resources/special-collections/guide-to-special-collections/at oz/womens-suffrage-move-ment-archives/ (accessed 3 August 2018).

Anon (2015), *Work–life Balance* [online]. Available from: https://www.mentalhealth.org.uk/a-to-z/w/work-life-balance (accessed 28 July 2018).

Anon (2013), 'How women view the world', BBC, [online], 23 October. Available from: https://www.bbc.co.uk/news/world-24583114 (accessed 27 June 2018).

Arcelus, J. et al. (2011), 'Mortality rates in patients with anorexia nervosa and other eating disorders: a meta-analysis of 36 studies', *Archives of General Psychiatry* [online], 68 (7), 724–31.

Aziz, Z. (2017), 'Gender dysphoria patients deserve better treatment than I can give them', *Guardian* [online], 15 August. Available from: http://www.theguardian.com/society/2017/

aug/15/gender-dysphoria-patients-need-specialists-not-gps (accessed 2 August 2018).

Bandelow, B. & Michaelis, S. (2015), 'Epidemiology of anxiety disorders in the 21st century', *Dialogues in Clinical Neuroscience*, 17 (3), 327–35.

Baxter, A. J. et al. (2013), 'Global prevalence of anxiety disorders: a systematic review and meta-regression', *Psychological Medicine* [online], 43 (5), 897–910.

Beebeejaun, Y. (2017), 'Gender, urban space, and the right to everyday life', *Journal of Urban Affairs* [online], 39 (3), 323–34.

Bliss, L. (n.d.), 'Finding the female flâneur' [online]. Available from: https://www.citylab.com/navigator/2017/03/when-women-walk-the-city/520542/ (accessed 27 June 2018).

Bondi, L. (1998), 'Gender, class, and urban space: public and private space in contemporary urban landscapes', *Urban Geography* [online], 19 (2), 160–85.

Borrow, A. P. & Handa, R. J. (2017), 'Estrogen receptors' modulation of anxiety-like behavior', *Vitamins and Hormones* [online], 10327–52.

Botteron, K. N. et al. (2002), 'Volumetric reduction in left subgenual prefrontal cortex in early onset depression', *Biological Psychiatry*, 51 (4), 342–4.

Bovens, L. & Marcoci, A. (2017), 'To those who oppose gender-neutral toilets: they're better for everybody', *Guardian* [online], 1 December. Available from: http://www.theguardian.com/commentisfree/2017/dec/01/gender-neutral-toilets-better-everybody-rage-latrine-trans-disabled (accessed 1 August 2018).

Brinton, R. D. et al. (2008), 'Progesterone receptors: form and function in brain', *Frontiers in Neuroendocrinology* [online], 29 (2), 313–39.

Bromet, E. et al. (2011), 'Cross-national epidemiology of DSM-IV major depressive episode', *BMC Medicine* [online], 990.

Bulik, C. M. et al. (1999), 'Fertility and reproduction in women with anorexia nervosa: a controlled study', *Journal of Clinical Psychiatry* [online], 60 (2), 130–35.

Bullivant, L. (2015), 'How are women changing our cities?', *Guardian* [online], 5 March. Available from: http://www.the guardian.com/cities/2015/mar/05/how-women-changing-cities-urbanistas-architecture-design (accessed 27 June 2018).

Butler, J. (1999), *Gender Trouble: Feminism and the Subversion of Identity*, London: Routledge.

——(n.d.), *Butler_Judith_Gender_trouble_feminism_and_the_subversion_of_identity_1990.pdf.* [online]. Available from: https://monoskop.org/images/e/ee/Butler_Judith_Gender_trouble_feminism_and_the_subversion_of_identity_1990.pdf (Accessed 2 August 2018).

Cafe, R. (2011), 'How the contraceptive pill changed Britain', *BBC News* [online], 4 December. Available from: https://www.bbc.com/news/uk-15984258 (accessed 3 August 2018).

Canu, E. et al. (2015), 'Brain structural abnormalities in patients with major depression with or without generalized anxiety disorder comorbidity', *Journal of Neurology* [online], 262 (5), 1255–65.

Castle, K. & Kreipe, R. (2007), 'Bulimia Nervosa', in Lynn C. Garfunkel et al. (eds.), *Pediatric Clinical Advisor (2nd Edition)* [online], 85–6. Available from: http://www.sciencedirect.com/science/article/pii/B9780323035064100483 (accessed 3 August 2018).

Celec, P. et al. (2015), 'On the effects of testosterone on brain behavioral functions', *Frontiers in Neuroscience* [online], 9. Available from: https://www.frontiersin.org/articles/10.3389/fnins.2015.00012/full (accessed 19 July 2018).

Chepesiuk, R. (2009), 'Missing the dark: health effects of light pollution', *Environmental Health Perspectives*, 117 (1), A20–A27.

Cohen, M. (2012), 'Why reproductive health is a civil rights issue', *Guardian* [online], 4 April. Available from: http://www.theguardian.com/commentisfree/cifamerica/2012/apr/04/reproductive-health-civil-rights-issue (accessed 27 June 2018).

Crenshaw, K. (n.d.), 'Demarginalizing the intersection of race and sex: a black feminist critique of antidiscrimination doctrine, feminist theory and antiracist politics', 31.

Dansky, B. S. et al. (2000), 'Comorbidity of bulimia nervosa and alcohol use disorders: results from the National Women's Study', *International Journal of Eating Disorders* [online], 27 (2), 180–90.

Dazed (2018a), 'New reality TV show Genderquake highlights the need for LGBTQ solidarity' [online]. Available from: http://www.dazeddigital.com/film-tv/article/39975/1/channel-4-genderquake-lgbt-solidarity-romario-markus (accessed 3 August 2018).

Delaney, B. (2018), 'The generation gap is back – but not as we know it', *Guardian* [online], 14 April. Available from: http://www.theguardian.com/commentisfree/2018/apr/14/the-generation-gap-is-back-but-not-as-we-know-it (accessed 2 August 2018).

Derntl, B. et al. (2009), 'Amygdala activity to fear and anger in healthy young males is associated with testosterone', *Psychoneuroendocrinology* [online], 34 (5), 687–93.

Ding, X.-X. et al. (2014), 'Maternal anxiety during pregnancy and adverse birth outcomes: a systematic review and meta-analysis of prospective cohort studies', *Journal of Affective Disorders* [online], 159103–10.

Doan, P. L. (n.d.), 'Queers in the American city: transgendered perceptions of urban space'.

—— (2016), 'The evolution of gay and lesbian spaces', in *Queerying Planning: Challenging Heteronormative Assumptions and Reframing Planning Practice*. Oxon: Routledge.

—— (2010), 'The tyranny of gendered spaces – reflections from beyond the gender dichotomy', *Gender, Place & Culture* [online], 17 (5), 635–54.

Drevets, W. C. et al. (1997), 'Subgenual prefrontal cortex abnormalities in mood disorders', *Nature* [online], 386 (6627), 824–7.

Dunkel Schetter, C. & Tanner, L. (2012a), 'Anxiety, depression and stress in pregnancy: implications for mothers, children, research, and practice', *Current Opinion in Psychiatry* [online], 25 (2), 141–8.

Durdiakova, J. et al. (2011), 'Testosterone and its metabolites – modulators of brain functions', *Acta Neurobiologiae Experimentalis*, 71 (4), 434–54.

Dutta, A. et al. (2014), 'Resting state networks in major depressive disorder', *Psychiatry Research: Neuroimaging* [online], 224 (3), 139–51.

Etkin, A. et al. (2009), 'Disrupted amygdalar subregion functional connectivity and evidence of a compensatory network in generalized anxiety disorder', *Archives of General Psychiatry* [online], 66 (12), 1361–72.

Fairburn, C. G. & Harrison, P. J. (2003), 'Eating disorders', *Lancet* [online], 361 (9355), 407–16.

Ferguson, L. & Harman, S. (2015), 'Gender and infrastructure in the World Bank', *Development Policy Review* [online], 33 (5), 653–71.

Fernandez-Aranda, F. et al. (2007), 'Symptom profile of major depressive disorder in women with eating disorders', *Australian & New Zealand Journal of Psychiatry* [online], 41 (1), 24–31.

Field, T. et al. (1990), 'Behavior-state matching and synchrony in mother–infant interactions of nondepressed versus depressed dyads', *Developmental Psychology* [online], 26 (1), 7–14.

Fleming, A. & Tranovich, A. (2016), 'Why aren't we designing cities that work for women, not just men?', *Guardian* [online], 13 October. Available from: http://www.theguardian.com/global-development-professionals-network/2016/oct/13/why-arent-we-designing-cities-that-work-for-women-not-just-men (accessed 27 June 2018).

Fowles, E. R. (1998), 'The relationship between maternal role attainment and postpartum depression', *Health Care for Women International* [online], 19 (1), 83–94.

Frank, G. K. et al. (2013), 'Alterations in brain structures related to taste reward circuitry in ill and recovered anorexia nervosa and in bulimia nervosa', *American Journal of Psychiatry* [online], 170 (10), 1152–60.

García-García, I. et al. (2013), 'Neural responses to visual food cues: insights from functional magnetic resonance imaging', *European Eating Disorders Review* [online], 21 (2), 89–98.

Garrett, M et al. [(2018), *Border Disruptions: Playbour & Transnationalisms* [online]. Available from: https://www.furtherfield.org/editorial-border-disruptions-playbour-transnationalisms/ (accessed 1 August 2018).

Ghayee, H. K. & Auchus, R. J. (2007), 'Basic concepts and recent developments in human steroid hormone biosynthesis', *Reviews in Endocrine & Metabolic Disorders* [online], 8 (4), 289–300.

Giltay, E. J. et al. (2012), 'Salivary testosterone: associations with depression, anxiety disorders, and antidepressant use in a large cohort study', *Journal of Psychosomatic Research* [online], 72 (3), 205–13.

Gonsalves, G. S. et al. (2015), 'Reducing sexual violence by increasing the supply of toilets in Khayelitsha, South Africa: a mathematical model', *PLOS ONE* [online], 10 (4), e0122244.

Gwynne, J. (2013), 'New femininity, neoliberalism, and young women's fashion blogs in Singapore and Malaysia', in *Bodies Without Borders*. New York: Palgrave Macmillan, 51–73. Available from: https://link.springer.com/chapter/10.1057/9781137365385_4 (accessed 24 July 2018).

Halberstam, J. (n.d.), *#29 Trans*: Bodies and Power in the Age of Transgenderism* [online]. Available from: http://www.documenta14.de/en/calendar/1016/-29-trans-bodies-and-power-in-the-age-of-transgenderism (accessed 27 June 2018). Halberstam, J. (2005), *In a Queer Time and Place: Transgender Bodies, Subcultural Lives*. London: New York University Press.

—— (2015), 'In/Human–Out/Human', *GLQ: A Journal of Lesbian and Gay Studies*, 21 (2), 239–42.

Hamilton, J. P. et al. (2012), 'Functional neuroimaging of major depressive disorder: a meta-analysis and new integration of baseline activation and neural response data', *American Journal of Psychiatry* [online], 169 (7), 693–703.

Hamson, D. K. et al. (2016), 'Sex hormones and cognition: neuroendocrine influences on memory and learning', *Comprehensive Physiology* [online], 1295–1337.

Hankin, B. L. et al. (1998), 'Development of depression from pre-adolescence to young adulthood: emerging gender differences in a 10-year longitudinal study', *Journal of Abnormal Psychology* [online], 107 (1), 128–40.

Hanman, N. (2013), 'Eve Kosofsky Sedgwick and Judith Butler showed me the transformative power of the word queer', *Guardian* [online], 22 August. Available from: http://www.theguardian.com/commentisfree/2013/aug/22/judith-butler-eve-sedgwick-queer (accessed 24 July 2018).

Haque, S. E. et al. (2014a), 'The effect of a school-based educational intervention on menstrual health: an intervention study among adolescent girls in Bangladesh', *BMJ Open* [online], 4 (7), e004607.

Harding, I. H. et al. (2015), 'Effective connectivity within the frontoparietal control network differentiates cognitive control and working memory', *NeuroImage* [online], 106144–53.

Harrison, P. (2018), 'Genderquake review – raw, funny, brittle and combative', *Guardian* [online], 7 May. Available from: http://www.theguardian.com/tv-and-radio/2018/may/07/gender-quake-review-reality-experiment-tv (accessed 3 August 2018).

Hayward, C. et al. (1997), 'Psychiatric risk associated with early puberty in adolescent girls', *Journal of the American Academy of Child & Adolescent Psychiatry*, 36 (2), 255–62.

Hepburn, M. (2005a), 'Social problems in pregnancy', *Anaesthesia & Intensive Care Medicine* [online], 6 (4), 125–6.

Herring, M. & Kaslow, N. J. (2002), 'Depression and attachment in families: a child-focused perspective', *Family Process*, 41 (3), 494–518.

Hickey, M. et al. (2012), 'Evaluation and management of depressive and anxiety symptoms in midlife', *Climacteric: The Journal of the International Menopause Society* [online], 15 (1), 3–9.

Hildebrandt, T. et al. (2010), 'Conceptualizing the role of estrogens and serotonin in the development and maintenance of bulimia nervosa', *Clinical Psychology Review* [online], 30 (6), 655–68.

Hinsliff, G. (2018), 'The Gender Recognition Act is controversial – can a path to common ground be found?', *Guardian* [online], 10 May. Available from: http://www.theguardian.com/world/2018/may/10/the-gender-recognition-act-is-controversial-can-a-path-to-common-ground-be-found (accessed 2 August 2018).

Holsen, L. M. et al. (2011), 'Stress response circuitry hypoactivation related to hormonal dysfunction in women with major depression', *Journal of Affective Disorders* [online], 131 (1–3), 379–87.

Hudson, J. I. et al. (2007), 'The prevalence and correlates of eating disorders in the National Comorbidity Survey replication', *Biological Psychiatry* [online], 61 (3), 348–58.

Jacobs, E. G. et al. (2015), '17β-estradiol differentially regulates stress circuitry activity in healthy and depressed women', *Neuropsychopharmacology* [online], 40 (3), 566–76.

Kahn, R. S. et al. (2005a), 'Intergenerational health disparities: socioeconomic status, women's health conditions, and child behavior problems', *Public Health Reports (Washington, D.C.: 1974)* [online], 120 (4), 399–408.

Kaiser, R. H. et al. (2015), 'Large-scale network dysfunction in major depressive disorder: a meta-analysis of resting-state functional connectivity', *JAMA Psychiatry* [online], 72 (6), 603–11.

Kessler, R. C. (2012), 'The costs of depression', *The Psychiatric Clinics of North America* [online], 35 (1), 1–14.

Kessler, R. C. et al. (2001), 'The epidemiology of generalized anxiety disorder', *The Psychiatric Clinics of North America*, 24 (1), 19–39.

Kessler, R. C. & Bromet, E. J. (2013), 'The epidemiology of depression across cultures', *Annual Review of Public Health* [online], 34 (1), 119–38.

Kim, K. R. et al. (2012), 'Functional and effective connectivity of anterior insula in anorexia nervosa and bulimia nervosa', *Neuroscience Letters* [online], 521 (2), 152–7.

Krishnan, V. & Nestler, E. J. (2008), 'The molecular neurobiology of depression', *Nature* [online], 455 (7215), 894–902.

Kristensen, J. et al. (2005a), 'Pre-pregnancy weight and the risk of stillbirth and neonatal death', *BJOG: An International Journal of Obstetrics & Gynaecology* [online], 112 (4), 403–8.

Lanphear, B. P. (2015), 'The impact of toxins on the developing brain', *Annual Review of Public Health* [online], 36 (1), 211–30.

Larson, Charles P. (2007), 'Poverty during pregnancy: its effects on child health outcomes', *Paediatrics & Child Health*, 12 (8), 673–7.

Lee, S. et al. (2014), 'Resting-state synchrony between anterior cingulate cortex and precuneus relates to body shape concern in anorexia nervosa and bulimia nervosa', *Psychiatry Research: Neuroimaging* [online], 221 (1), 43–8.

Lewis, Gemma et al. (2018), 'The association between pubertal status and depressive symptoms and diagnoses in adolescent females: a population-based cohort study', *PLOS ONE* [online], 13 (6), e0198804.

Maeng, L. Y. & Milad, M. R. (2015a), 'Sex differences in anxiety disorders: interactions between fear, stress, and gonadal hormones', *Hormones and Behavior* [online] 76106–17.

Maron, E. et al. (2018), 'Imaging and genetic approaches to inform biomarkers for anxiety disorders, obsessive-compulsive disorders, and PTSD', *Current Topics in Behavioral Neurosciences* [online].

Martin, E. I. et al. (2010), 'The neurobiology of anxiety disorders: brain imaging, genetics, and psychoneuroendocrinology', *Clinics in Laboratory Medicine* [online], 30 (4), 865–91.

McDonald, H. et al. (2018), 'Ireland votes by landslide to legalise abortion', *Guardian* [online], 26 May. Available from: http://www.theguardian.com/world/2018/may/26/ireland-votes-by-landslide-to-legalise-abortion (accessed 3 August 2018).

McFarland, J. et al. (2011), 'Major depressive disorder during pregnancy and emotional attachment to the fetus', *Archives of Women's Mental Health* [online], 14 (5), 425–34.

McHenry, J. et al. (2014), 'Sex differences in anxiety and depression: role of testosterone', *Frontiers in Neuroendocrinology* [online], 35 (1), 42–57.

McLean, C. P. et al. (2011), 'Gender differences in anxiety disorders: prevalence, course of illness, comorbidity and burden of illness', *Journal of Psychiatric Research* [online], 45 (8), 1027–35.

Mettler, L. N. et al. (2013), 'White matter integrity is reduced in bulimia nervosa', *International Journal of Eating Disorders* [online], 46 (3), 264–73.

Micali, N. et al. (2007), 'Risk of major adverse perinatal outcomes in women with eating disorders', *British Journal of Psychiatry* [online], 190 (3), 255–9.

Misri, S. et al. (2015), 'Perinatal generalized anxiety disorder: assessment and treatment', *Journal of Women's Health* [online], 24 (9), 762–70.

Moraga-Amaro, R. et al. (2018), 'Sex steroid hormones and brain function: PET imaging as a tool for research', *Journal of Neuroendocrinology* [online], 30 (2), e12565.

Murray, L. et al. (1996), 'The impact of postnatal depression and associated adversity on early mother–infant interactions and later infant outcome', *Child Development*, 67 (5), 2512–26.

Nestler, E. J. et al. (2002), 'Neurobiology of depression', *Neuron* [online], 34 (1), 13–25.

Nestler, E. J. & Carlezon, W. A. (2006), 'The mesolimbic dopamine reward circuit in depression', *Biological Psychiatry* [online], 59 (12), 1151–9.

Newcomb, R. (2006), 'Gendering the city, gendering the nation: contesting urban space in Fes, Morocco, *City & Society* [online], 18 (2), 288–311.

Ongley, H. (2016), 'The subversive power of hyper-feminine fashion', *i-D* [online]. Available from: https://i-d.vice.com/en_us/article/zmx558/the-subversive-power-of-hyper-feminine-fashion (accessed 1 August 2018).

Otte, C. et al. (2016), 'Major depressive disorder', *Nature Reviews Disease Primers* [online], 216065.

Parker, G. B. & Brotchie, H. L. (2004), 'From diathesis to dimorphism: the biology of gender differences in depression', *Journal of Nervous and Mental Disease*, 192 (3), 210–16.

Pigott, T. A. (2003), 'Anxiety disorders in women', *The Psychiatric Clinics of North America*, 26 (3), 621–72, vi–vii.

Pmhdev (n.d.), *Follicular Phase*, National Library of Medicine [online]. Available from: https://www.ncbi.nlm.nih.gov/pubmed health/PMHT0024716/ (accessed 22 July 2018).

Preiser, J.-C. et al. (2016), 'Successive phases of the metabolic response to stress', in [online], 5–18.

Quinn, V. P. et al. (2017), 'Cohort profile: Study of Transition, Outcomes and Gender (STRONG) to assess health status of transgender people', *BMJ Open* [online], 7 (12). Available from: https://www.ncbi.nlm.nih.gov/pmc/articles/PMC5770907/ (accessed 3 August 2018).

Rajkowska, G. et al. (1999), 'Morphometric evidence for neuronal and glial prefrontal cell pathology in major depression', *Biological Psychiatry* [online], 45 (9), 1085–98. See also accompanying Editorial in this issue.

Reed, C. (1996), 'Imminent domain: queer space in the built environment', *Art Journal* [online], 55 (4), 64–70.

Regitz-Zagrosek, V. (n.d.), 'Sex and gender differences in health', *EMBO Reports* [online], 13 (7), 596–603.

Robichaud, M. & Debonnel, G. (2005), 'Estrogen and testosterone modulate the firing activity of dorsal raphe nucleus serotonergic neurones in both male and female rats', *Journal of Neuroendocrinology* [online], 17 (3), 179–85.

Ross, L. E. & McLean, L. M. (2006), 'Anxiety disorders during pregnancy and the postpartum period: a systematic review', *Journal of Clinical Psychiatry*, 67 (8), 1285–98.

Salbach-Andrae, H. et al. (2008), 'Psychiatric comorbidities among female adolescents with anorexia nervosa', *Child Psychiatry and Human Development* [online], 39 (3), 261–72.

Sangiuliano, M. (n.d.), 'Smart cities and gender: main arguments and dimensions for a promising research and policy development area', 9.

Scharfe, E. (2007), 'Cause or consequence? Exploring causal links between attachment and depression', *Journal of Social and Clinical Psychology* [online], 26 (9), 1048–64.

Schmid, H. (2018) [online]. Available from: https://www.helgaschmid.com/ (accessed 3 August 2018).

—— (2016): [online]. Available from: http://uchronia.world (accessed 3 August 2018).

Schüle, C. et al. (2011), 'Neuroactive steroids in affective disorders: target for novel antidepressant or anxiolytic drugs?', *Neuroscience* [online], 19155–77.

Shansky, R. M. (2009), 'Estrogen, stress and the brain: progress toward unraveling gender discrepancies in major depressive disorder', *Expert Review of Neurotherapeutics* [online], 9 (7), 967–73.

Shansky, R. M. & Arnsten, A. F. T. (2006), 'Estrogen enhances stress-induced prefrontal cortex dysfunction: relevance to major depressive disorder in women', *Dialogues in Clinical Neuroscience*, 8 (4), 478–81.

Sheldon, R. (2016), *Queer Universal* [online]. Available from: https://www.e-flux.com/journal/73/60456/queer-universal/ (accessed 19 July 2018).

Shin, L. M. & Liberzon, I. (2010a), 'The neurocircuitry of fear, stress, and anxiety disorders', *Neuropsychopharmacology* [online], 35 (1), 169–91.

—— (2010b), 'The neurocircuitry of fear, stress, and anxiety disorders', *Neuropsychopharmacology* [online], 35 (1), 169–91.

Shrestha, P. et al. (n.d.), 'Layer 2/3 pyramidal cells in the medial prefrontal cortex moderate stress-induced depressive behaviors', *eLife* [online], 4. Available from: https://www.ncbi.nlm.nih.gov/pmc/articles/PMC4566133/ (accessed 1 August 2018).

Slowik, A. et al. (2018), 'Brain inflammasomes in stroke and depressive disorders: regulation by estrogen', *Journal of Neuroendocrinology* [online], 30 (2).

Solnit, R. (2018), 'Feminists have slowly shifted power. There's no going back', *Guardian*, [online], 8 March. Available from: http://www.theguardian.com/commentisfree/2018/mar/08/feminists-power-metoo-timesup-rebecca-solnit (accessed 27 June 2018).

Solomon, M. B. & Herman, J. P. (2009), 'Sex differences in psychopathology: of gonads, adrenals and mental illness', *Physiology & Behavior* [online], 97 (2), 250–58.

Spade, D. (n.d.), *Toward a critical trans politics* [online]. Available from: http://www.deanspade.net/wp-content/uploads/2013/03/upping-the-anti-interview-2012.pdf (accessed 1 August 2018).

Spira-Cohen, A. et al. (2011), 'Personal exposures to traffic-related air pollution and acute respiratory health among Bronx school children with asthma', *Environmental Health Perspectives* [online], 119 (4), 559–65.

Stamets, P. (2012), '*Trametes versicolor* (turkey tail mushrooms) and the treatment of breast cancer', *Global Advances in Health and Medicine* [online], 1(5), 20. Available from: https://doi.org/10.7453/gahmj.2012.1.5.007

Stein, A. et al. (2012), 'Maternal cognitions and mother–infant interaction in postnatal depression and generalized anxiety disorder', *Journal of Abnormal Psychology* [online], 121 (4), 795–809.

Stein, D. J. et al. (2017), 'Epidemiology of anxiety disorders: from surveys to nosology and back', *Dialogues in Clinical Neuroscience*, 19 (2), 127–36.

Stein, M. B. & Sareen, J. (2015), 'Clinical practice: generalized anxiety disorder', *New England Journal of Medicine* [online], 373 (21), 2059–68.

Stephens, M. A., & Wand, G. (2012), 'Stress and the HPA axis: role of glucocorticoids in alcohol dependence', *Alcohol Research: Current Reviews*, 34(4), 468–83.

Stock, K. (2018a), 'Changing the concept of "woman" will cause unintended harms', *The Economist* [online], 6 July. Available from: https://www.economist.com/open-future/2018/07/06/changing-the-concept-of-woman-will-cause-unintended-harms.

—— (2018b), 'Notes for my talk to A Woman's Place UK, Brighton, 16th July 2018', Medium [online]. Available from: https://medium.com/@kathleenstock/notes-for-my-talk-to-a-womans-place-uk-brighton-17th-july-2018-f1 b607414119 (accessed 2 August 2018).

Swinbourne, J. M. & Touyz, S. W. (2007), 'The co-morbidity of eating disorders and anxiety disorders: a review', *European Eating Disorders Review* [online], 15 (4), 253–74.

Teti, D. M. et al. (n.d.), 'Maternal depression and the quality of early attachment: an examination of infants, preschoolers, and their mothers', 13.

Titova, O. E. et al. (2013), 'Anorexia nervosa is linked to reduced brain structure in reward and somatosensory regions: a meta-analysis of VBM studies', *BMC Psychiatry* [online], 13 (1), 110.

Topping, A. (2018), 'Debate over inclusion of trans women in women-only spaces intensifies', *Guardian* [online], 9 February.

Available from: http://www.theguardian.com/world/2018/feb/09/debate-over-inclusion-of-trans-women-in-women-only-spaces-intensifies (accessed 2 August 2018).

UC Berkeley Events (n.d.), *Gayatri Chakravorty Spivak on Situating Feminism* [online]. Available from: https://www.youtube.com/watch?reload=9&v=garPdV7U3fQ (accessed 27 June 2018).

Uher, R. et al. (2014), 'Major depressive disorder in Dsm-5: implications for clinical practice and research of changes from Dsm-Iv', *Depression and Anxiety* [online], 31 (6), 459–71.

Venator, R. V. R. & J. (2001), 'Gender gaps in relative mobility', Brookings [online]. Available from: https://www.brookings.edu/blog/social-mobility-memos/2013/11/12/gender-gaps-in-relative-mobility/ (accessed 31 July 2018).

Vesga-López, O. et al. (2008), 'Psychiatric disorders in pregnant and postpartum women in the United States', *Archives of General Psychiatry* [online], 65 (7), 805–15.

Vivian-Taylor, J. & Hickey, M. (2014a), 'Menopause and depression: is there a link?', *Maturitas* [online], 79 (2), 142–6.

Walf, A. A. & Frye, C. A. (2006), 'A review and update of mechanisms of estrogen in the hippocampus and amygdala for anxiety and depression behavior', *Neuropsychopharmacology* [online], 31 (6), 1097–1111.

Wang, P. S. et al. (2007), 'Worldwide use of mental health services for anxiety, mood, and substance disorders: results from 17 countries in the WHO World Mental Health (WMH) surveys', *Lancet* [online], 370 (9590), 841–50.

Watson, H. J. et al. (2015), 'Psychosocial factors associated with bulimia nervosa during pregnancy: an internal validation study', *International Journal of Eating Disorders* [online], 48 (6), 654–62.

Weiss, N. et al. (2009), 'The blood–brain barrier in brain homeo-stasis and neurological diseases', *Biochimica et Biophysica Acta (BBA) – Biomembranes* [online], 1788 (4), 842–57.

Wharton, W. et al. (2012), 'Neurobiological underpinnings of the estrogen–mood relationship', *Current Psychiatry Reviews* [online], 8 (3), 247–56.

Whitmore, M. J. et al. (2014), 'Generalized anxiety disorder and social anxiety disorder in youth: are they distinguishable?', *Child Psychiatry & Human Development* [online], 45 (4), 456–63.

Wilkins, R. et al. (1991a), 'Birth outcomes and infant mortality by income in urban Canada, 1986', *Health Reports*, 3 (1), 7–31.

Wingen, G. A. van et al. (2009), 'Testosterone increases amyg-dala reactivity in middle-aged women to a young adulthood level', *Neuropsychopharmacology* [online], 34 (3), 539–47.

Yang, C.-F. J. et al. (2007), 'Testosterone levels and mental rotation performance in Chinese men', *Hormones and Behavior* [online], 51 (3), 373–8.

Yilmaz, Z. et al. (2015), 'Genetics and epigenetics of eating disorders', *Advances in Genomics and Genetics* [online], 5131–50.

Zeidan, M. A. et al. (2011), 'Estradiol modulates medial prefrontal cortex and amygdala activity during fear extinction in women and female rats', *Biological Psychiatry* [online], 70 (10), 920–27.

Zipfel, S., Giel, Katrin E., et al. (2015), 'Anorexia nervosa: aetiology, assessment, and treatment', *Lancet Psychiatry* [online], 2 (12), 1099–111.

ACKNOWLEDGEMENTS

I want to dedicate this book to everyone involved in it, quoted in it and those who were brave and generous in sharing their stories too. I also need to thank my editor Celia Hayley who kept me firmly on track and inspired throughout, along with my agent Charlie Brotherstone and incredible publisher Mireille Harper. Without them this book wouldn't have happened, and it was truly a team effort.

Lola Ross, nutritionist and co-founder of Moody

Karla Vitrone, COO and co-founder of Moody

Minisha Sood, endocrinologist

Tara Swart, neuroscientist

Araceli Camargo, neuroscientist and researcher

Melissa Hemsley, chef

Saschan Fearon-Josephs, founder of the Womb Room

Sharmadean Reid, founder of BeautyStack

Kenny Ethan Jones, activist and model

Shara Tochia, co-founder of Whatever Your Dose

Farah Kabir, co-founder of HANX

Sarah Welsh, co-founder of HANX and gynaecologist

Ellamae Fullalove, founder of Va Va Womb and Mind Over MRKH

Shardi Nahavandi, co-founder of Pexxi

Cheyenne Morgan, nurse and founder of Let's Talk Gynae

Hannah Winkler, activist and coder

Susan Masters, head of nursing at RCN

Lola Adesioye, activist and journalist

Claire Gorham, teacher

I also owe huge thanks to my dearest friends, family and team who were cheerleading me on throughout the writing process.